D0619987

BUILD YOUR OWN HOME LAB

By

Clement S. Pepper

BUILD YOUR OWN HOME LAB

By

Clement S. Pepper

A Division of Howard W. Sams & Company
A Bell Atlantic Company

PROMPT© Publications is an imprint of Howard W. Sams & Company, A Bell Atlantic Company, 2647 Waterfront Parkway, E. Dr., Indianapolis, IN 46214-2041.

International Standard Book Number: 0-7906-1108-2

Acquisitions Editor: Candace M. Hall
Editor: Natalie F. Harris
Assistant Editors: Pat Brady, Loretta Leisure
Editorial Support: Karen Mittelstadt, Shey Query
Typesetting: Natalie Harris
Indexing: Natalie Harris
Cover Design: Christy Pierce
Graphics Conversion: Bill Skinner, Terry Varvel, Ted Wilson III
Illustrations: Courtesy of the Author
Additional Illustrations and Materials: Courtesy of Analog Devices, Harris Semiconductor Corporation, Motorola Semiconductor Products, National Semiconductor Corporation, Popular Electronics Magazine, Purdy Electronics/Interoptics, QT Optoelectronics, Teccor Electronics, Inc., Temic Semiconductors. Product illustrations and product information in this book are the property of their respective manufacturers and are used here with permission.

PRINTED IN THE UNITED STATES OF AMERICA

9 8 7 6 5 4 3 2 1

CONTENTS

To my family: Lana, Ronald, and David, who contributed more than they can ever know to the making of this book.

FOREWORD

FOREWORD

Between the years of 1960 and 1990, my engineering employment was with R & D and one-of-a-kind system developments. It was also in these years that the incredible expansion of solid state technology took place — the transition from simple (so to speak) discretes to the present day integrated circuit devices of almost unbelievable complexity.

I was fortunate in having supportive employers who enabled me to keep abreast of new developments as they became available. The semiconductor manufacturers were very active in promoting their product lines, with seminars and dinner meetings from which I returned with arm loads of the latest data books. I found myself working with cohorts with whom, in our daily contacts, I traded knowledge and experience.

My employers were supportive of continuing education, and over the years I took a number of courses — I obtained a BS in Physics, Electronics Option, from San Diego State University in 1963 while working at the Scripps Institution of Oceanography in La Jolla, California.

Even so, I realized there was no way I could learn all I needed and wanted to know within the confines of my daily work on the job or in long, drawn-out university courses. I would have to find a way to make more use of my own time, for that which could not be done on the job.

Hence the home lab.

In the long run, guess where much of what I put into my job originated?

Simply put, my purpose in preparing this book is to encourage a career in the field of electronics among young men and women. Twenty years ago, the computer captured the attention of many of us in its construction. We were hobbyists and experimenters with as much interest in the computer's internal parts and functions as with its usage. It is no longer appropriate for most of us as individuals to construct or even attempt to fathom the ins and outs of our own computers from the bare bones. However, the computer is an electronic system. It is within our capability, and to our advantage, to understand the workings of the parts and circuits that comprise the construction and operation of the tools we need, from test equipment to computers. However, first and foremost, this is a book to help with building a career. If we learn how to build the basic tools we need for our home electronics lab, then others will follow.

WHAT'S INSIDE

Chapter 1: A Look Back At the Future

The author indulges in a bit of autobiography. Thoughts on looking ahead from a long-ago vantage point. The young man or woman of today is hopefully viewing his or her own prospects for the future in an increasingly challenging technical world.

Chapter 2: Getting Started

Not a cookbook, this chapter is an initializer for your idea generator. Our home lab will not spring into being at the snap of a clip lead. We need to keep our thoughts on goals — we should think to ourselves, where do *I* want to go and what is *my* best way to get there? What is essential for the here and now, and what can be fitted in as time goes by?

Chapter 3: An Instrumentation Base of Our Own Making

Perhaps it seems a bit odd to have the end result right up front; but why not enjoy the dessert first and use it to hone our appetite for the filler courses? In this chapter are instrument projects we can do for ourselves, complete with circuit descriptions, parts lists, and construction details. Certain to contribute to the functioning of our lab, to say nothing of our satisfaction in achievement.

Chapter 4: Breadboards and Prototype Modules

A number of manufacturers are offering materials that contribute to the ease and efficiency in constructing a working prototype of our project. When we find a recurring need for certain parts or circuits, why not put together basic, reusable modules of them? A bit of time well spent saves more time for when it counts.

Chapter 5: Circuit Data Basics For Us To Build Upon

We can relate this chapter to the menu in our favorite restaurant. It shows the good things to choose from; but to actually taste the reality of it, we have to place an order. Taken to the limit, this chapter could fill several books. The range of devices from simple discrete diodes to more elegant A/D converters will provide the insight on what is most suitable for our project. The devices selected for each of the chapter's four sections are those the author believes to be of general interest. Source references for further information are liberally provided.

Appendices

What it is and where it is to be found: catalogs, databooks, and surplus part sources. These are included with thought for those of us who must rely on the telephone and UPS[1] for much of our information and material needs.

ACKNOWLEDGMENTS

An author does not produce his masterpiece in an isolation cell. Along the trail that begins with the first faltering sentence, to the typing of that one final character, innumerable others are contributing from their backpacks of encouragement, advice and assistance. Many thanks to these individuals.

---

1. *Trademark, United Parcel Service.*

CHAPTER 1

A LOOK BACK AT THE FUTURE

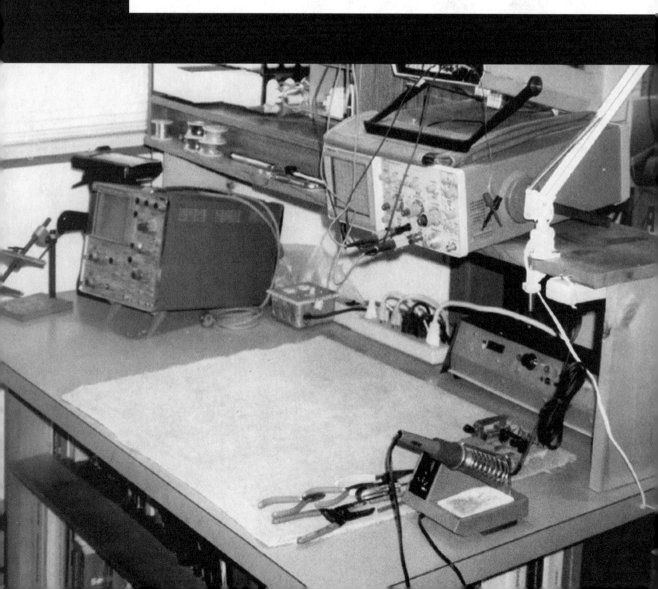

CHAPTER 1
A LOOK BACK AT THE FUTURE

It was my twentieth birthday. I was standing on the deck of the USS Harwood, a destroyer operating off the coast of China in the spring of 1946. I remember as though it were yesterday — telling myself I had fifty years to make something of my life.

Those fifty years are up now, and looking back I believe I have made something of my time. Much of what I have accomplished I feel I owe to the early set up and continuing development of an electronics capability at my home for over forty of those years; not as a business, though there were a couple brief forays, nor just as a hobby. I simply wanted the freedom and the ability to pursue interests and learning that would not be available to me as an employee following the dictates of an employer.

I am a "hands on" person, and I would expect that is true of yourself also. There is no doubt in my mind that the ability to explore and learn in our own way on electronic topics of interest to ourselves makes a valuable contribution to a career in any technical capacity. It is my reason for this writing — if you are a young man or woman wondering what the future holds in store for you, as was I a half century ago, perhaps something here will offer encouragement.

Throughout the mid 1950s, I assembled a number of kits from the Heathkit company. These included an oscilloscope, a vacuum tube voltmeter, a vacuum tube tester, and an audio generator. I passed the word around that I would work on television sets at no charge other than for material costs.

My objective was to learn — and did I ever! There were many invitations for dinner, with reminders to "bring your tube tester."

In the mid 1960s, I acquired the notion of writing articles for publication in the electronic publications of the time: chiefly *Radio & TV News, Radio-Electronics,* and *Popular Electronics.* This turned out to be an excellent financial aid for the continuing development of my lab.

The growth of my career and that of my home enterprise were proceeding hand-in-hand. As time passed, I developed the lab to the point where I could do much of my employer's design work more efficiently at home. With management blessing, this is what I did for many of the projects I worked on. This was a real benefit in that the project was providing the materials; since I carried many parts from one project to the next, I built up a substantial stock with which to work.

A questionable advantage was the ability to work my own hours — frequently well into the night and many weekends. There is a downside to this: it can disrupt your family life, and I urge some restraint in this regard. Save time for your loved ones.

Now let's consider what the designation *home lab* really stands for.

Basically, it is a place where learning can take place. All else is detail — work space, equipment, parts, library — all of it. Granted, these are necessary to a working facility; but the real essential is the desire to seek, learn and develop the strengths and interests that we have.

My lab began years ago, on a side table in my bedroom. I moved it to the garage (once I had one), and eventually, to the one room in the house that was my own for just that purpose. At present, I live in an apartment: the lab consists of one bedroom, a couple of closets, and on occasion, the kitchen table. More on this in the next chapter. It's compact, but you will be impressed at how much can be fit in.

A lab is not cheap — it can cost a lot of money, as a matter of fact. The trick is to try to not do everything all at once, and to look for ways to make it pay its own way. My way, in part, was article writing. Also, San Diego, my home for many years, hosted a wide variety of surplus shops and I haunted them all. Much of my parts stock was obtained by disassembling used equipment.

Much of what we need by way of instrumentation we can make for ourselves. Chapter 3 leads the way on this. Also, there is much available as reconditioned equipment — 100 MHz oscilloscopes that cost over $2000 are available for $500. Just be a little careful.

I am now retired from outside employment, but my home-based activities continue. These have included consulting, the design and construction of a unique measurement system, a challenging software task to automate a motion picture theatre, technical writing, and a host of activities just for enjoyment.

In the beginning of this section, I spoke of my hope that something you, the reader, find in this book will provide encouragement for your own future in electronics. Much of the current interest among younger people seems to be with the computer — let us not forget that the computer is an electronic device. A very complex device to be sure; but we should keep in mind that the most complex system is still the sum of its parts. If we understand the basics, the system will follow. Also, let's not brush "analog" aside as outmoded. Perish the thought: all of nature is analog. The A/D converter will always be with us.

One parting thought: in the early 1950s I took a correspondence course from the Capitol Radio Engineering Institute (CREI). Their motto was, "The man who knows how will always have a job; the man who knows why will be his boss."

Think about it while planning your own home lab.

CHAPTER 2

GETTING STARTED

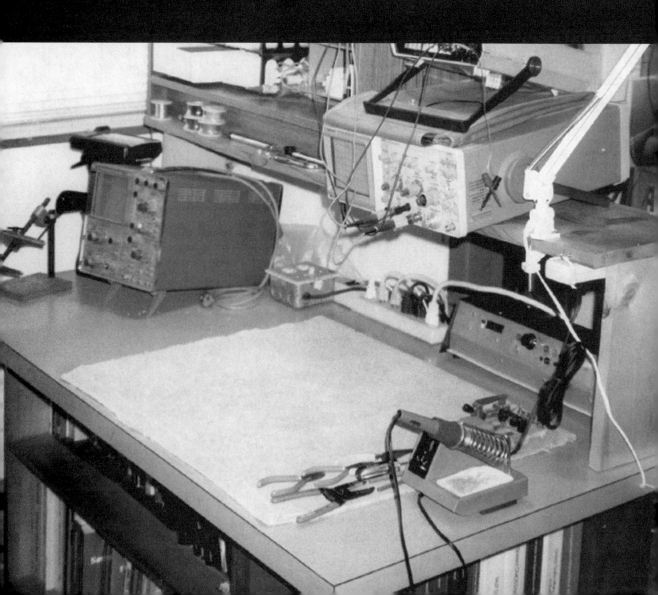

CHAPTER 2
GETTING STARTED

The intent of this book is to present ideas — food for thought, to coin a phrase. This is my task. Yours is to compare what I have to offer with your needs. There will never be a 100 percent match, because we both know that just because something has worked for one is no guarantee for another. Still, if a tidbit or two triggers an "ah ha!" on your part, the mission has been accomplished.

2.1. GOALS AND DECISIONS: WHAT ARE WE SHOOTING FOR?

You have a reason for creating an island of electronic expertise in your home: your own personal objectives. We know the field of electronics is a broad one, encompassing an extensive range of specialties. Still, there does exist within the diversity an underlying commonality. In this as well as the chapters following, keep a weather eye out for that which falls in your area of interest. Just be sure you know what that area is. Keep it well in mind.

If you don't find a fit somewhere in here, I haven't done my job.

2.2. SPACE: FOR DOING WHAT, AND HOW MUCH?

You can see the layout of my own home lab in *Figure 2.1*. It's not all that big a room, measuring 12.7 by 10.7 feet, leaving out the closet. As you can see, I have packed a lot into it, which goes for the closet as well. In years past, my work area has ranged up and down, from a folding table in the corner of my bedroom to one whole room plus an attic, plus a garage. It's not the square footage that matters; it's what we make of it. The old vacuum tube days when instruments were big, bulky and heavy are long gone.

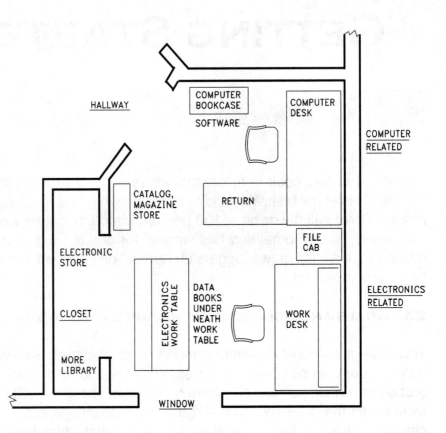

Figure 2.1. *Layout of the author's home lab. Room size is 11 by 13 feet — still, a lot has been fit in. Some tasks such as chassis cutting and drilling usually end up on the kitchen table.*

Present-day integrated circuits allow us to do ever so much more in a compactly-arranged environment. I do find myself walking sideways at times just to get through.

There are three kinds of space: floor, work, and storage. We think of floor space as so many square feet to move around in. The work space is where things get done: desk and bench tops.

Storage is more than where we keep the goodies to be used in our projects. This is where we squirrel away our books, tools, instruments, floppy disks,

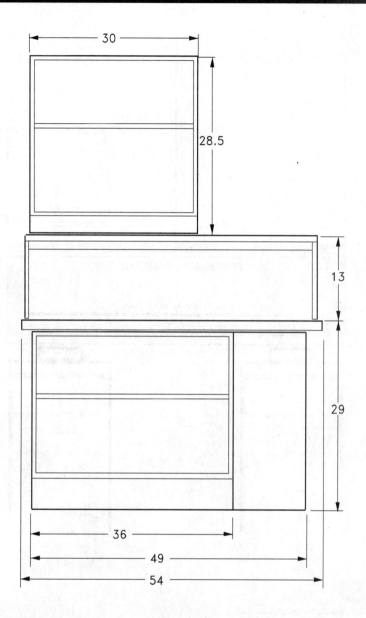

Figure 2.2. *There is very little wasted space in this vertically organized work bench and equipment "tower complex." The bookcase was assembled from a kit.*

and those odds and ends that have a way of getting underfoot. Storage may be neat and well organized. It may also be chaotic. Most labs seem to have a mix of both.

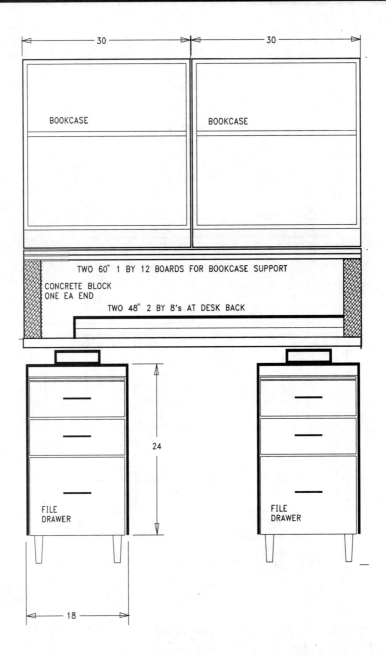

Figure 2.3. This work desk arrangement places needed reference material readily at hand. The two 48" planks lift books above desktop clutter. Lattice "windows" in the concrete blocks keep the weight down.

Photo 2.1*. Work desk organization. Note the work bench corner. A swivel chair, omitted for clarity, allows quick access to either work surface.*

2.3. FURNISHINGS: SURFACES AND STORAGE

The core of my electronics work center consists of two back-to-back work surfaces: a desk and a bench. These are vertically organized towers as you can see in **Figures 2.2** and **2.3**, and also in **Photos 2.1** and **2.2**. I like the data, tools and parts I work with to be conveniently close at hand, and this arrangement works perfectly in that respect.

The desk has the dimensions of your typical office item. A couple of 2" x 8" planks across the back supports the data books I like to have at hand. The elevating planks make access easier with the top cluttered, as it usually is. Concrete blocks are solid, and they don't shift easily. The bookcases were assembled from kits.

I had the bench top made to order some 25 years ago. It rests on two bookcases. These hold the majority of my semiconductor data books. A couple of cardboard boxes laid on their sides behind the two bookcases provide out-of-sight storage.

The shelf across the back supports an oscilloscope and digital voltmeter on the right; a bookcase similar to those on the desk at the left. Stacked *in/out* trays offer space for breadboards and the small modules I frequently construct.

2.4. BASIC TOOLS: PLIERS TO SOLDER STATIONS

As with so much of life, it's not the quantity of tools that counts so much as their utility. After all, how many needlenose pliers can we use at one time? For years I made do with a cheapy, one-size-strips-all wire stripper. Finally I wised up and invested in one that is suitable for the range of wires commonly used: 30 to 22 gauge. Similarly, I also invested in a proper soldering station versus a simple iron, though I keep a high wattage iron on hand for the heavier tasks that come up. Solder wick is good, but often a solder puller is even better when I have to make circuit board changes. **Table 2.1** is a listing of the basic tools I have found to be the most useful.

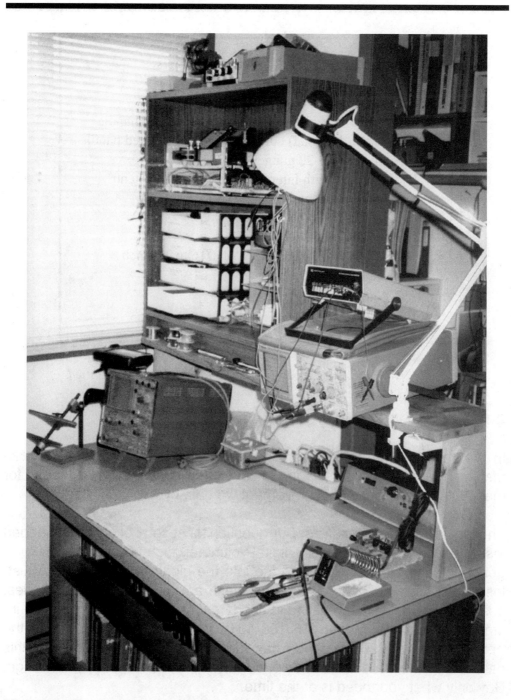

Photo 2.2*. The work bench arrangement. The two supporting bookcases contain a portion of the authors's data book library. The bookcase at the left on the shelf provides storage for breadboards and prototype modules.*

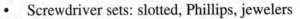

- Screwdriver sets: slotted, Phillips, jewelers
- Trimmer potentiometer adjustment tool
- Nut driver set
- 4-1/2" long nose pliers, serrated tip, insulated handle
- 4-1/2" side (diagonal) cutter, flush cut, insulated handle
- Wire stripper, 30-22 gauge
- Soldering iron with interchangeable tips and stand
- Vacuum desoldering bulb
- Desoldering braid
- X-Acto knife
- Electric hand drill with chuck of 1/16 to 1/4 inch
- 1/2" reamer
- Drill bit set, 1/16 — 1/4"
- Hex wrench set

Table 2.1. Some useful items to have on hand when getting started.

2.5. INSTRUMENTATION: MAKE OR BUY

In the chapters that follow, we will learn of instrumentation that we can make for ourselves. This is in part to save a few coins; but in larger part for the learning and sense of achievement we gain by it.

There is, of course, no *make* or *buy* question for several of our most needed instruments. The oscilloscope, digital voltmeter, high frequency waveform generator and counters that operate into the gigahertz region are "must" instruments for purchase. However, these are all available at a modest cost from many sources dealing in used, reconditioned equipment. For instance, a Tektronix Model 465 100 MHz scope, originally costing in the neighborhood of $2000, can be had for as little as $400. Scopes of this caliber are advertised in the trade publications in the $400-$600 price range. Buy only what your need is at the time.

On the other hand, much low-frequency analog and digital test, measurement, and control instrumentation is readily adaptable to our own home construction. Examples of these are described later in this book. The cost of these do-it-yourself projects is partly offset by whatever you have squirreled away in your parts storage, plus the availability of much material on the surplus and hobby sales offerings. It pays to shop around.

2.6. PARTS STOCK: WHAT TO HAVE ON HAND, AND HOW MANY

Back in the 1960s and '70s, I responded to mail-order bargain offers and haunted surplus stores for what was mostly obsolete equipment discarded for more up-to-date devices. After a time, I realized I was working too much with past technology, all in the desire to save on cost. I did salvage lots of useful passive components and hardware, but I no longer pull semiconductors from discarded circuit boards.

I try to maintain a fairly decent stock of the parts I typically use. I live in a area where the only nearby outlet is Radio Shack. While they have an excellent stock of certain parts, there is much they do not carry, and I have to rely on catalog purchasing. There is nothing so aggravating as discovering the need for a critical part in the middle of a project. If your pattern is like mine, I must admit that once started, like the Energizer Bunny, I just want to keep on going and going until it's done.

2.7. LIBRARY: DATA AND CATALOG SOURCES

Up-to-date information is essential in these days of rapid technological change. We cannot all be engineers but keeping informed in our particular areas of interest is vital to continued learning. Current mail-order catalogs and many semiconductor data books are available for the asking. Many companies provide application notes on their products that are informative on a broader basis. Subscriptions to magazines such as *Popular Electronics* helps us keep informed. Browse through the Appendices of this book to see what is there to meet your needs.

2.8. ORGANIZING AND PLANNING: KEEPING TRACK

Years ago while working on a very important assignment, I discovered an individual working from a pencil drawing spread out on the floor. This doesn't sound too terribly bad, except it was the original drawing, with no copies and already boasting a couple of footprints. If our project is worth an out-pouring of time and money, it is only reasonable to protect its value with adequate documentation. For home projects, a simple lab notebook or 3-ring binder will do for holding our plans.

2.9. SOME DEARLY HELD BELIEFS

1. Maintain a neat, orderly work area. A neat desk is not necessarily the sign of a disordered mind.

2. Document the progress of your design genius as you go along. It saves a

    lot of wear and tear on our memory cells later.

3. Check your wiring; double check as you go.

4. Insert an ON/OFF switch between your power supply and the circuit you are working with. Frequent line switching can have bad long-term effects on the supply.

5. Be open to change. First ideas are seldom the best.

6. Take it easy. Haste does, in truth, often lead to waste.

7. Learn good soldering techniques. Poor soldering is the source of more malfunctions than design errors.

8. Jot down your thoughts before they vaporize.

9. Use sockets for your integrated circuits.

10. Keep the receipts and invoices that come with your parts in a folder or 3-ring binder, for ready reference.

No doubt you have favorite beliefs of your own to contribute to the list.

CHAPTER 3

AN INSTRUMENTATION BASE OF OUR OWN MAKING

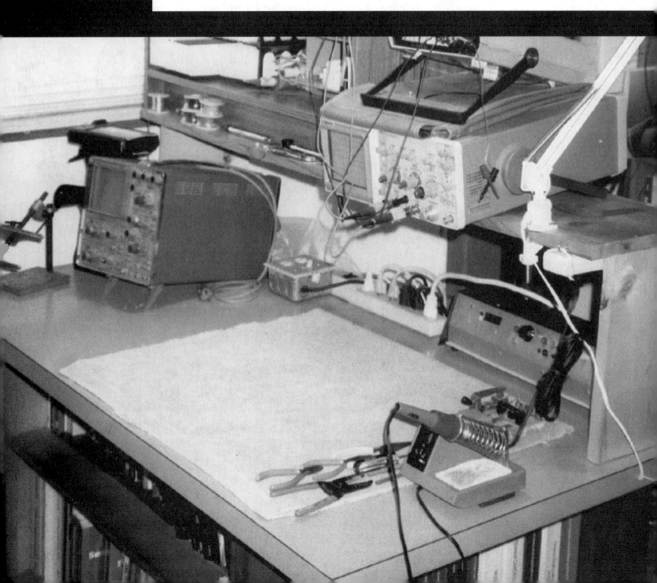

CHAPTER 3
AN INSTRUMENTATION BASE OF OUR OWN MAKING

There is a surprising amount of very useful equipment we can design and construct on our own. In general, commercially equivalent items will have more features, many of which we may have little need for, while adding to the cost. What I am describing in this chapter arose in response to needs that came up for benchtop linear power supplies, precision DC voltage and current sources, and waveform generators. I make no claim that these are comparable to more sophisticated professional instrumentation; but they perform, they meet the need, and there is the satisfaction of doing it *my way*.

The performance obtained depends not only on the design, but how well it is implemented in the construction and testing. Take time to be careful. Each circuit has been constructed and its performance verified. All needed circuit and component information is provided. Although my parts mounting and housing is shown for some items, the final choice is however you want it. The modules described in Chapter 4 may provide some helpful ideas.

Some Thoughts on Case Cutouts and Drilling

Through experience I have evolved a few rules for myself on housing preparation. All the circuits shown here may be installed in either plastic or aluminum cases. For the most part, I prefer aluminum boxes with sloping faces for ease of readability.

I begin with the cutout that is most challenging. Getting the hardest out of the way first makes the remaining seem easier.

I start with a template of the hole pattern, using the blank side of pasteboard of the sort found in cereal boxes. Beware of taping over painted surfaces; unprimed metal boxes are prone to peeling the paint along with the tape. Rubber cement does not present this problem; I use it freely, for both templates and labels.

I debur holes using a reamer or larger diameter drill bit. For rectangular cutouts, I drill a series of small holes around the perimeter, leaving about 1/32" margin for safety. A small flat file does the job well. Depending on the cutout, a nibbler can also be used to advantage. Black paper cutouts affixed with rubber cement make attractive bezels while covering over the rough edges.

Standard dual banana plug spacing is 3/4". Jacks should be spaced accordingly.

Hold down the clutter. I've seen bench and desktops that resembled a landfill. A bit of cleanup every so often will save time and stress later when you search for the thingamajig that you "just know is in here somewhere."

You may not think so now, but at the end you'll be showing your handiwork to everyone from the love of your life to the mail carrier. Beyond that, try to make a good looking job of it, if for no other reason than the sheer satisfaction.

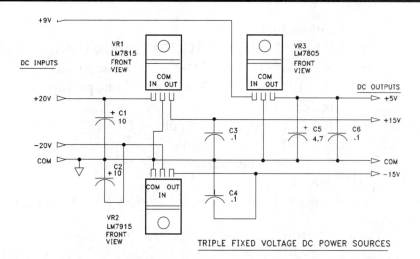

Figure 3.1.1. *A regulator circuit for three independent outputs of +5V, and +/- 15 volts. These can be constructed as a unit or as two independent supplies for bench service.*

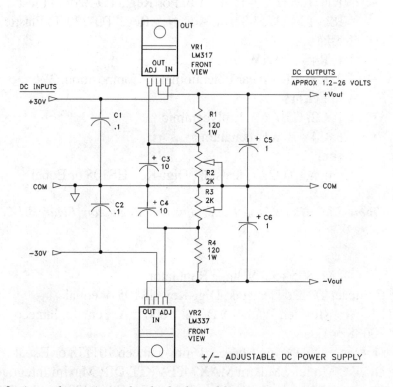

Figure 3.1.2. *A regulator circuit for dual adjustable outputs over the range of about 1.3 to 26 volts. When combined with the DVM project of section 3.2 a versatile bench supply with digital readout is achieved.*

Semiconductors
- VR1 - LM7815, +15V, TO-220(T) plastic package
- VR2 - LM7915, -15V, TO-220(T) plastic package
- VR3 - LM7805, +5V, TO-220(T) plastic package

Capacitors
- C1, C2 - 10 mF, 25V Tantalum
- C3, C4, C6 - .1 mF, Ceramic
- C5 - 4.7 mF, Tantalum

Other
- Compact TO220 Heat Sink, Digi-Key HS105 or Equal

Table 3.1.1. Parts List - Triple Output Regulator, Figure 3.1-1.

Semiconductors
- VR1 - LM317, 3-Term. Adj. Pos Reg., TO-220(T) Plastic
- VR2 - LM337, 3-Term. Adj. Neg Reg., TO-220(T) Plastic

Resistors
- R1, R4 - 120, 1W, 5%
- R2, R3 - 2K Linear Potentiometer, Composition, 2W

Capacitors
- C1, C2, C5, C6 - .1 mF ceramic
- C3, C4 - 10 mF, tantalum

Other
- Compact TO220 Heatsink, Digi-Key HS105 or Equal

Table 3.1.2. Parts List - Dual Adjustable Regulator, Figure 3.1-2.

- VR1 - LM7805 +5V Voltage Regulator
- Compact TO220 Heatsink, Digi-Key HS105 or equal
- P1 - Transformer, Wall, 9 VDC @ 1000 mA, Female, Jameco
- 10088 or Equal
- J1 - Jack, DC Power, Male, 2.1 mm, Jameco 101178 or Equal
- DC/DC Module, Maxim MAX743EVKIT-DIP, Maxim Integrated
- Products, 120 San Gabriel Dr., Sunnyvale, CA 94086, (408) 737-7600

Table 3.1.3. Parts List - DC/DC Converter, Figure 3.1-3.

3.1. LINEAR, FIXED AND VARIABLE VOLTAGE, DC POWER SUPPLIES

Back in my not-so-well-to-do days of the 1960s, I discovered that high-voltage power transformers could be hooked up in reverse for low-voltage transistor use. However, in those days, these transformers were widely available as surplus and practically given away. I don't recommend doing it now.

In practice, many designers opt for the exciting tasks, leaving the power design till last — leading to many late hours and lost weekends. The designs here are suitable for bench supplies and for the constructions that follow. The assortment of low-cost, high-performance, voltage regulators now available has taken the drudgery out of power supply design and worked wonders for reliability.

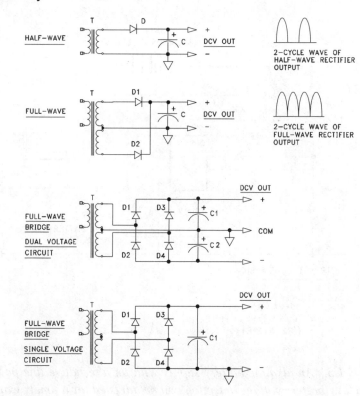

Figure 3.1.3. Four frequently used transformer/rectifier circuits with a capacitor input filter.

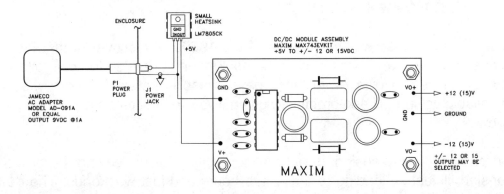

Figure 3.1.4. *This DC/DC converter is an elegant solution to the need for dual supplies when a five volt supply is available.*

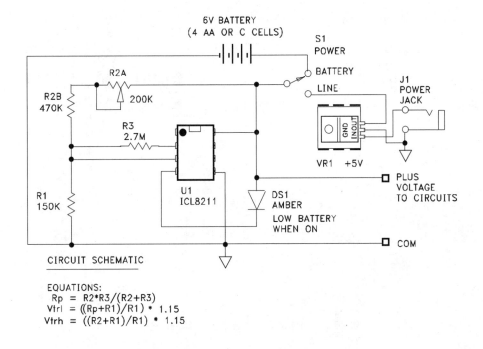

EQUATIONS:
Rp = R2*R3/(R2+R3)
Vtrl = ((Rp+R1)/R1) * 1.15
Vtrh = ((R2+R1)/R1) * 1.15

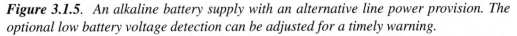

Figure 3.1.5. *An alkaline battery supply with an alternative line power provision. The optional low battery voltage detection can be adjusted for a timely warning.*

The circuits of *Figures 3.1.1* and *3.1.2* are the basis for my bench power supply, in service since 1980 with no failures. It consists of a +5V supply at 300 mA, a dual $\pm$ 15V at about 200 mA, and a dual adjustable $\pm$ 1.3 to $\pm$ 26V at about 200 mA. This last is provided with the DVM described in the next section. The DVM is switched between the positive and negative outputs at a one second rate, with a red LED identifying the positive.

At the time I constructed my supply, the metal cased LM309, LM320, and LM340 were in vogue. If desired, they can be used in place of the LM7805, LM7915, and LM7815 shown. The primary difference is a slightly higher wattage rating. The LM78XX series in general can deliver up to one ampere with adequate heatsinking.

The source of dissipation in the regulator arises from the current flow and the voltage difference between the input and output. The regulators typically require a minimum of two volts differential for proper regulation, so we need to select rectifier components that hold down the differential.

Heat sinks are cheap. It is almost impossible to select one that is too big. A dry surface at the interface is a real no-no. Use a thermal compound. Hot spots in a thermal path are analogous to IR drops in an electrical circuit. The rise in temperature is the product of thermal resistance and heat flow. The heat flow is unavoidable, but containable.

- U1 - ICL8211 Programmable Voltage Detector, Harris
- Semiconductor or Equal
- VR1 - LM7805 +5V Voltage Regulator
- R1 - Resistor, 5%, 1/4W, 150K
- R2B - Resistor, 5%, 1/4W, 470K
- R3 - Resistor, 5%, 1/4W, 2.7M
- R2A - Potentiometer, Trimmer, Cermet, 200K Ohms
- DS1 - Amber LED
- Batteries, 4 Each, AA or C Size Cells
- Compact TO220 Heatsink, Digi-Key HS105 or Equal

Table 3.1.4. Parts List - Alkaline Battery Power Source, Figure 3.1-5.

Thoughtful transformer selection is essential. It is this, even more than the regulator, that will determine the available power from your supply. With proper heatsinks the regulators shown are good for a full ampere.

The two figures pretty much speak for themselves; there is very little to be added. These were adapted from application examples in the *National Semiconductor Linear Devices* data book[1].

The Transformer/Rectifier

Selection of a rectifier circuit, shown in **Figure 3.1.3**, is more of a challenge. The four configurations shown will cover every power supply need for our home lab.

Interpreting transformer specs is the tough part. What is meant by a secondary rating of 12.6V at 250 mA? Alternating current follows a sinusoidal wave ranging from zero to a peak value. Ratings are expressed as the Root-Mean-Square (RMS) which is .707 times the peak value. The voltage peak is then 1/.707 • VRMS, or 1.41 • VRMS. 12.6 • 1.41 = 17.8. Ideally, our filter capacitor should charge to this peak value. In reality, losses in the transformer winding and the diode voltage drop(s) will reduce this by two or three volts or more, depending on the transformer and the load current. Also, we must keep in mind that the filter draws current in spurts, further reducing the peak.

None of the projects which follow depend on a transformer/rectifier circuit. Their power is derived from a wall unit or alkaline batteries. This greatly simplifies the task. If you construct a power supply for the bench I strongly recommend two transformers, one for each of the two figures. This is because the variable supply operates at almost double the voltage of the fixed. A full-wave, two-diode circuit with 1000 mF filters will do the job.

A DC/DC Converter Option

Figure 3.1.4 illustrates an attractive dual power resolution where only a 5-volt supply is available, or is there for other purposes[2]. The converter load

1. *National Semiconductor Corporation, General Purpose Linear Devices Databook, 1989*

must be light — not greater than 100 mA. That requirement is met by the two projects in this chapter which make use of it (Sections 3.3 and 3.4).

The Alkaline Battery Option

Three of the projects that follow use this approach to their power needs. Batteries are suitable where the demand is low and/or the use is intermittent with a low duty cycle. Sometimes it is advisable to include a line power backup, as is illustrated in *Figure 3.1.5*.

An optional low battery voltage detection feature is also shown. I typically set R2A for a 4.3 volt detection.

3.2. A UTILITY DVM; A DC POWER SUPPLY DVM

When it comes to readability, a comparison between an analog meter and a digital display is no contest at all. The two DVM circuits in this section can be constructed for about the same cost as a good quality analog meter.

The first circuit leads to a useful, general purpose DC digital voltmeter. The second is great for displaying the output values of a dual, variable bench power supply.

3.2.1. A Utility DVM

The absolute value amplifier is easily constructed and provides excellent results. Just how excellent, of course, will depend on the quality of the parts used, the care taken in its construction, and in the calibration. With a fifty-volt input range and a four-digit display, the resolution is to the nearest ten millivolts.

The schematic also includes a voltage-to-frequency converter circuit. I typically combine construction and testing. In this instance, I suggest constructing and calibrating the amplifier before proceeding with the converter.

---

2. *Maxim Integrated Products, Dual-Output, Switch-Mode Regulator, MAX743EVKIT Power Supply Evaluation Kit.*

The amplifier design is based on a concept by George A. Philbrick[1]. He had vacuum tubes in mind then; but it works just as well, probably better, with solid state. *Figure 3.2.1* illustrates the concept.

The uniqueness of the circuit is in the two diodes in the amplifier feedback loop. When **ein** is negative, the amplifier output is positive relative to the input by the drop of diode D2. This cuts off diode D1. The following amplifier inverts the negative **ein**. When **ein** is positive, the amplifier A1 is negative at the output, enabling diode D1 while cutting off D2. The negative output is then coupled to A2 by resistor R4. Resistors R1-R3 and R5 are equal value while R4 is one-half this.

We see that the gain of A2 is unity with **ein** negative. This because with the inverting input of A1 clamped at **+ein** the net voltage in this path is zero.

With **ein** positive the output of A2, is:

$$eo = 2ein - ein = +ein$$

— because of the gain of 2 in the lower path.

The role of amplifier A1 is seen as steering the positive or negative **ein** to A2 in a manner to always obtain a positive value for its output, **eo**.

In the circuit of *Figure 3.2.2* A1 and A2 are seen as U1B and U1C, two sections of a LM324. U1A is an input stage to provide scaling of the source, **ein**.

We take advantage of the output of U1B to inform us of a positive input. The output of comparator U2 goes low with a positive input, turning on the LED.

The LM331 converter does not like a negative input. Therefore, it is wise to verify the amplifier functioning before continuing with the converter.

The initial step in the calibration is to connect a known negative (-ein) voltage to the input. If the LED glows, you are in trouble. Adjust input trimmer

1. *George A. Philbrick Researches, Inc. Applications Manual For Computing Amplifiers For Modelling Measuring Manipulating & Much Else, 1996, p 59.*

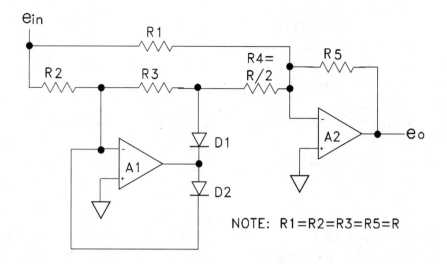

Figure 3.2.1. *The absolute value amplifier function. The diodes at the first stage output make the polarity decision.*

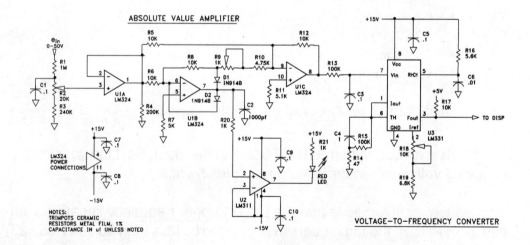

Figure 3.2.2. *An absolute value amplifier accepts DC input voltages of either polarity from zero to 50 volts. The circuit also includes a voltage-to-frequency converter.*

Semiconductors
- U1 - LM324 Quad Op Amp
- U2 - LM311 Analog Comparator
- U3 - LM331 Precision Voltage-to-Frequency Converter
- D1, D2 - 1N914B Diode or Equal
- DS1 - Red LED

Resistors - 1/4W, 1% Metal Film unless otherwise noted
- R1 - 1M
- R2 - 20K Ceramic Trimpot
- R3 - 240K
- R4 - 200K
- R5, R6, R8, R10, R12, R17 - 10K
- R7, R11 - 5.1K
- R9 - 1K Ceramic Trimpot
- R13, R15 - 100K
- R14 - 47
- R16 - 5.6K
- R18 - 10K Ceramic Trimpot
- R19 - 6.8K
- R20, R21 - 1K

Capacitors - Ceramic
- C1, C3, C5, C7-C10 - .1 mF
- C2 - 1000 pF
- C4 - 1 mF
- C6 - .01 mF

Table 3.2.1. Parts List - The Absolute Value Amplifier, Figure 3.2-2.

R2 to obtain an output from U1C of one-fifth the input. So if the your input is minus 25 volts, your meter should read plus 5 volts.

The reason for this ratio is that the LM331 upper frequency linearity is limited to 10 KHz. If the maximum **ein** is going to be less than 50 volts (say 25) then revise the ratio accordingly.

Next, replace the negative **ein** with a positive of equal value. Now adjust trimmer R9 to obtain the same output from U1C.

Semiconductors
- U1, U2 - LM741 Op Amp
- U3 - CD4066 CMOS Quad Analog Switch
- U4 - LM331 Precision Voltage-to-Frequency Converter
- U5 - ICL7555 CMOS Timer
- U6 - CD4013 CMOS Dual D Flip-Flop
- Q1 - 2n3704 or Equal NPN Transistor
- DS1 - Red LED

Resistors - 1/4W, 1% Metal Film Unless Otherwise Noted
- R1, R3 - 20K
- R2, R4, R6, R8 - 10K
- R5 - 10K Ceramic Trimpot
- R7 - 470K
- R9 - 4.7K, 5%
- R10, R17 - 10K, 5%
- R11 - 10K Ceramic Trimpot
- R12 - 6.8K
- R13, R15 - 100K
- R14 - 47
- R16 - 5.6K
- R18 - 1K, 5%

Capacitors - Ceramic
- C1, C6 - .01 mF
- C2, C4 - 1 mF
- C3, C5, C7-C12 - .1 mF

Table 3.2.2. Parts List - A Dual Power Supply DVM, Figure 3.2-3.

That's it for the amplifier.

Calibrate the converter by adjusting the amplifier for a positive ten volt output. Now if all is as it should be, adjusting trimmer R18 for a frequency of 10 KHz should do it.

3.2.2. A DC Power Supply DVM

The circuit of *Figure 3.2.3* is one I built into my bench supply way back in 1980. It is still doing its job very well. It really makes a difference when adjusting the voltages.

Because we are going to be continuously monitoring a positive and a negative input we have to take a different approach; no absolute value amplifier here.

For a positive input, a simple noninverting buffer, U1 (a LM741), will do nicely. A positive input requires inversion. To equal the positive in magnitude a feedback network is needed. Adjust R5 for the same amplifier outputs with equal magnitudes of plus and minus input voltages.

We use two of the four channels of a CD4066 to route our voltages to the converter. Note there is no change in the converter circuit from the previous DVM. A third CD4066 channel switches the LED to ON for the positive input.

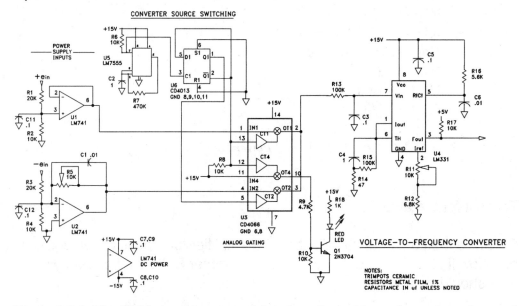

Figure 3.2.3. *This DVM circuit is very useful to show the positive and negative voltages of a dual power supply on a digital display. The switching circuit alternates the display values at about a one second rate.*

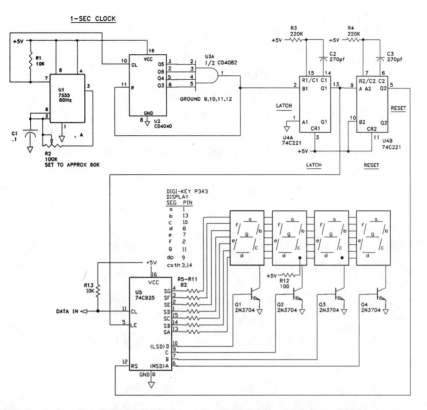

Figure 3.2.4. *A one-second clock/divider circuit and four-digit LED display for use with either of the two DVM circuits.*

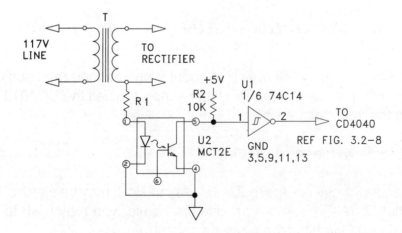

Figure 3.2.5. *Using the 60Hz line as the clock source.*

Semiconductors
- U1 - ICL7555 CMOS Timer
- U2 - CD4040 CMOS 12-Bit Binary Counter
- U3 - CD4082 CMOS Quad AND Gate
- U4 - 74C221 CMOS Dual Monostable
- U5 - 74C925 CMOS Decade Counter/Decoder/Driver
- Q1-Q4 - 7-Segment Common Cathode LED Display

Resistors - 5% Composition Unless Noted Otherwise
- R1 - 10K
- R2 - 100K Ceramic Trimpot
- R3, R4 - 220K
- R5-R11 - 82
- R12 - 100

Capacitors - Ceramic
- C1 - .1 mF
- C2, C3 - 270 pF

Table 3.2.3. Parts List - 1-Second Clock Divider and Display, Figure 3.2-4

- U1 - 74C14 CMOS Schmitt Hex Inverter
- U2 - MCT2E or Equal Optical Coupler
- T - Line Power Transformer
- R1 - As required for 10 -20 mA Coupler Current
- R2 - 10K, 5%

Table 3.2.4. Parts List - 60 Hz Line Clock Divider, Figure 3.2-5

The 7555 timer values will switch at about a one-second rate; sufficient for a good reading of the display. The timer simply toggles the CD4013 flip-flop to select the input for viewing.

3.2.3. Clock Divider and Display

The 1-second circuit of *Figure 3.2.4* is very basic. It may be used, of course, with either DVM. For a more precise time base, you may wish to go to a more elaborate circuit using a quartz crystal.

While I show a 7555 timer source, another alternative is to apply the low voltage output of your power supply's line transformer to an optical coupler. This is shown in **Figure 3.2.5**. The coupler output is best buffered by a 74C14 Schmitt input inverter.

3.3. A PRECISION DC CONSTANT VOLTAGE/CURRENT SOURCE

Some years back, Intersil developed a line of integrated circuits as "engineering solutions on a chip." I first used their ICL8211 in the early 1980s as a battery low-voltage detector, and found it to be a low-cost, high-performance device. I have since learned that it and a companion device, the ICL8212, are two ICs possessing properties of value to many applications.

The Intersil product line is now offered by Harris Semiconductor[1]. Some details of these truly versatile guys are provided in Chapter 5.2.7.

3.3.1. A High-Performance Voltage and Constant Current Source

I had need for a highly stable source of DC millivolt level potentials and microampere circuits. The impressive regulation of the ICL8212 is the basis of the design of this experimenter's precision instrument of many uses. **Figure 3.3.1** shows the overall circuitry.

The Voltage Source

This design provides continuously variable voltages over the ranges of 0 to 1.00 volts and 0 to 10.00 volts simultaneously. The LF355 op-amp in the non-inverting mode is a good choice for this. The amplifier is bipolar with a JFET input featuring a low bias current. **Figure 5.2.25** illustrates the measurement circuit I used to assess the ICL8212 performance as a precision reference. In this circuit, the reference obtained from the ICL8212 is adjusted to precisely one volt at the top of potentiometers R5 and R8.

Advantage is taken of the high input impedance of the non-inverting unity gain amplifier configuration. A gain of ten stage, amplifier A3, is driven by the output of follower A1.

1. *Harris Semiconductor Corporation, Linear ICs for Commercial Applications, 1990. p 2-147, ICL8211/ ICL8212 Programmable Voltage Detectors.*

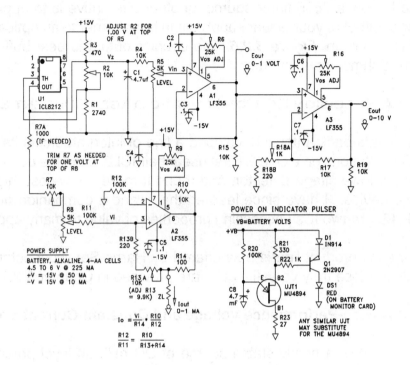

Figure 3.3.1. *The voltage and current source circuit schematic.*

To zero the amplifier, use an ohmmeter to approximately center the R6 potentiometer wiper. Short the two inputs to each other and to ground. Now slowly work the wiper of the adjusting trim pot to obtain a zero output. Repeat with A3 and trimmer R16.

The Current Source

It is highly desirable to reference the current load to ground. A circuit that meets this requirement was designed by B. Howland[2, 3], and has been associated with his name as the Howland Current Pump. The circuit is shown in **Figure 3.3.2**.

2. *Philbrick Researches, Inc., Applications Manual for Computing Amplifiers for Modelling Measuring Manipulating & Much Else, 1966, p. 66, III.6 Current Pump IV - Grounded Source, Grounded Load.*
3. *Thomas Frederiksen, Intuitive IC Op Amps, Linear Design Group, National Semiconductor Corporation, Santa Clara, CA.*

Semiconductors
- A1-A3 - LF355 Low Power JFET Input Op Amp
- U1 - ICL8212 Programmable Voltage Detector
- UJT1 - MU4894 Unijunction Transistor, Motorola or Equal
- D1 - 1N914 Diode or Equal
- Q1 - 2N2907 or Equal PNP Transistor
- DS1 - Red LED

Resistors 1% Metal Film Recommended Otherwise Noted
- R1 - 2740 Ohm
- R2, R7, R13A - 10K Ceramic Trimpot
- R3 - 470 Ohm
- R4, R10, R15, R17, R19 - 10K
- R5, R8 - 5K 10T Precision WW Pot., Digi-Key 73JB502 or Equal
- R6, R9, R16 - 25K Ceramic Trimpot
- R7A - 1K
- R11, R12, R14 - 100K
- R13B, R18B - 220 Ohm
- R14 - 100 Ohm
- R18A - 1K Ceramic Trimpot
- R20 - 100K, 5%
- R21 - 330 Ohm, 5%
- R22 - 1K, 5%
- R23 - 27 Ohm, 5%

Capacitors - Ceramic Unless Otherwise Noted
- C1, C8 - 4.7 mF, Tantalum
- C2-C7 - .1 mF

Table 3.3.1. Parts List - The voltage and current source circuit schematic, Figure 3.3-1

Referring again to **Figure 3.3.1**, note that 1.00 volt is to be present at the top of potentiometer R8. It is possible that the tops of R5 and R8 could have simply been tied together; but I preferred to maintain their isolation as a precaution.

Zero amplifier A2 as with A1 and A3.

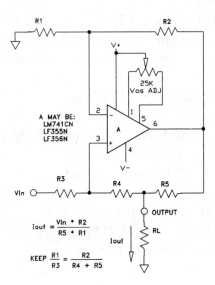

Figure 3.3.2. The Howland Current Pump circuit.

The resistor relationships in this circuit are critical. For this reason, the circuit provides a 10K cermet trim pot with a 220 ohm resistor in series, so as to assure the exact adjustment required — namely, to obtain a precise current range from 0 to 1.00 mA with a load variation of zero to 10K ohms. To obtain this, the amplifier zero and the adjustment of R13 must be carefully done. Some interaction occurs. Repeat the trimming until the same 1.00 mA current flows with the 10K load and the load shortened to ground.

Power Supply

The possibility of 60 Hz interference, ground loops and a desire for portability led me to opt for battery operation. *Figure 3.1.5* describes a low battery voltage detection circuit. The circuit as an assembly on a small circuit board of its own is described as a module in Chapter 4.2.14. *Figure 3.3.1* includes an optional UJT *power on* indicator pulser. It may be convenient to locate its parts on the same small board.

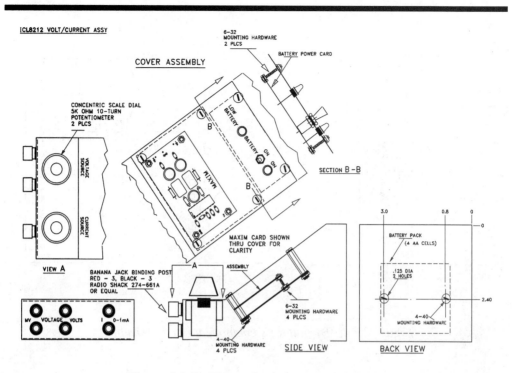

Figure 3.3.3. *The author's housing assembly.*

- Banana Jack Binding Post, Black - 3 each
- Banana Jack Binding Post, Red - 3 each
- Concentric Scale Dial, 10 Turns, Digi-Key 461KL or Equal
- Case - LMB 007 446 (Gateway Electronics Carries)
- Battery, AA Cell - 4 Each
- Battery Holder
- Mounting Hardware
- Selected Items from Table 3.1-3, Alkaline Battery
- Power Source

Table 3.3.2. *Parts List - The author's housing assembly, Figure 3.3-3.*

Note the need for a DC/DC converter for the $\pm$ 15 volt power. The MAXIM MAX743EVKIT[4] is an excellent choice. I do suggest using a line powered supply for the circuit checkout; then converting to the DC/DC converter as a final step.

4. Maxim Integrated Products, Dual-Output, Switch-Mode Regulator, MAX743EVKIT Power Supply Evaluation Kit.

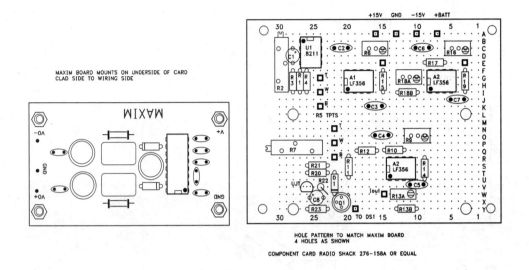

Figure 3.3.4. The author's component parts assembly.

The observed battery loading is 225 mA. Four AA cells with a constant fixed load of 225 mA dropped almost exactly one volt in three hours. With intermittent loading, this would be improved upon. In the load test, the set point of 4.5 volts occurred at six hours.

Photo 3.3.1. The assembled voltage/current source.

Construction

Figure 3.3.3 shows the author's housing. I chose this particular case for its sloping face and a compactness that saves on benchtop space. There is no need to follow my example; the assembly is shown as a generator for your own ideas.

The same applies to the component assembly shown in **Figure 3.3.4**. With this, or any assembly scheme, it pays to complete the wiring and verify the performance before installing it all in the housing. **Photo 3.3.1** shows the completed voltage/current source.

Applications

This instrument has value wherever a precision measurement is required. I make use of the voltage source with DC amplifier design, with DVM calibrations, and analog voltage meters, for example. A current source is excellent for DVM and analog meter testing, measuring zener voltages at various currents, and in circuit designs requiring a constant current input. The Howland circuit is of special value for data transmission over long lines, in that both the source and load are referenced to ground.

3.4. A SQUARE-TRIANGLE-SINE WAVEFORM GENERATOR

Hobbyists, circuit designers, service technicians — all find a multitude of uses for a versatile function generator. It is indispensable with any kind of audio, acoustic, passive or active filters, and ultrasound experiments, to name a few.

The ICL8038 Waveform Generator

The Harris[1] ICL8038 precision waveform generator is a monolithic integrated circuit capable of generating a variety of waveforms over an extended range of frequencies. We're using it here for its triangle, square, and sine wave provision over a frequency range of a few Hertz to 100 KHz. You can learn more about it in Chapter 5, section 2.6.

---

1. *Harris Semiconductor Corporation, Linear ICs For Commercial Applications, 1990. p 8-51, ICL8038 Precision Waveform Generator/Voltage Controlled Oscillator.*

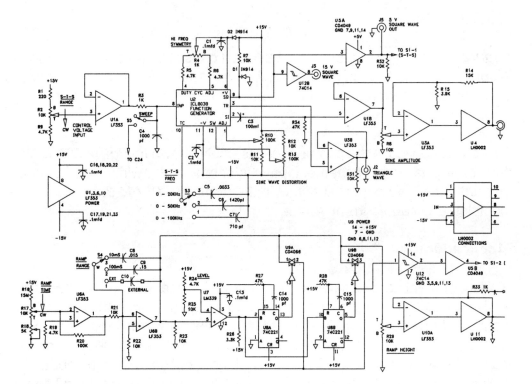

Figure 3.4.1. *Square-triangle-sine wave formation is shown in the top half of the schematic; the independent ramp generation is in the lower.*

Waveform Generator Features

The author's instrument produces simultaneous square, triangle, and sine waves (S-T-S). The square wave is brought out at both a 15V and 5V level as a working convenience. An independently generated linear sawtooth ramp is also included. **Figure 3.4.1** is the circuit schematic for these.

Five waveform outputs are provided via BNC panel jacks: the 5 and 15 volt square waves, the triangle and sine waves, and the ramp. All outputs are buffered. The sine and ramp outputs have output level controls. Precision ten-turn wire-wound potentiometers with turn-counting dials provide frequency and ramp time selection. The BNC jacks are on the housing back side to keep the connecting cables out of the way.

Semiconductors

- U1, U3, U6, U10 - LF353N Dual Op Amp
- U2 - ICL8038CCPD Function Generator, Harris
- U4, U11 - LH0002CN Current Amplifier, National
- U5 - CD4049B Hex Buffer
- U7 - LM339N Quad Comparator
- U8 - 74C221N Dual Monostable
- U9 - CD4066B Quad Analog Switch
- U12 - 74C14N Hex Inverting Schmitt Trigger
- D1 - 1N914 or Equal

Resistors: 1/4W, 5% Unless Otherwise Noted

- R1 - 220 Ohm
- R2, R17 - Potentiometer, WW, 10 Turn, 10K
- Digi-Key P/N 73JB10K, 2 Required
- R3, R33 - 1K
- R5, R6, R9, R19, R24 - 4.7K
- R7, R11, R12, R21, R22, R23, R25, R30, R31, R32 - 10K
- R10, R13 - 100K Ceramic Trimpot
- R14 - 15K
- R15 - 3.9K
- R16 - 15M
- R18 - 5K Ceramic Trimpot
- R20 - 100K
- R26 - 3.3K
- R27, R28, R34 - 47K

Capacitors: Low Voltage Miniature Ceramic Unless Otherwise Noted

- C1, C2, C11-C13, C16-C23 - .1 mF, Ceramic
- C3 - 100 mF, Tantalum
- C4 - —
- C14, C15 - 1000 pF, Ceramic
- C5 - .0033 mF Mylar
- C6 - 1420 pF, Mylar
- C7 - 710 pF, Mylar or Mica
- C8 - .015 mF, Mylar
- C9 - .5 mF, Mylar
- C10 - See Text

Table 3.4.1. Parts List - Square-Triangle-Sine Waveform Generator Circuit, Figure 3.4-1.

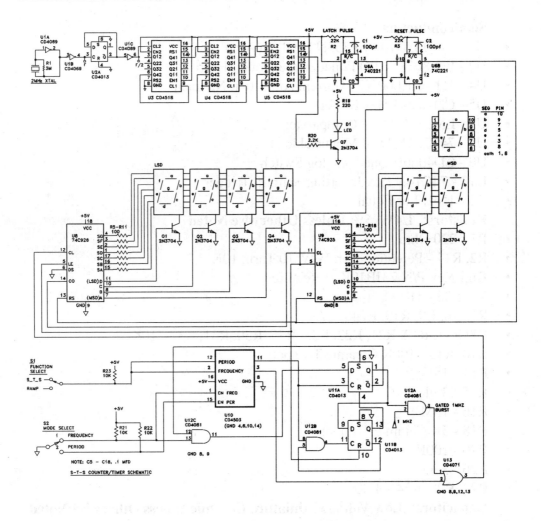

Figure 3.4.2. The counter/timer circuit schematic.

Square-Triangle-Sine Wave Generation

The upper half of the schematic relates to the S-T-S. A wire-wound, ten-turn potentiometer, R2, coupled with a precision ten-turn counting dial is used to set the frequency. The available range is determined by the capacitance at the three positions of rotary switch S3. R1 sets the lower frequency end, R9 the upper. A JFET input amplifier, U1A, buffers the input and provides a low impedance drive for the ICL8038. The 1K resistor, R3, is there to enable the ramp input for swept frequency use.

Semiconductors
- Q1-Q7 - Transistor, NPN, 2N3704 or Equal
- D1 - Red LED, MV5020 or Equal w/Mounting
- U1 - CD4069B/74C04 Hex Inverter
- U2, U11 - CD4013B Dual D-Flip-Flop
- U3-U5 - CD4518B Dual BCD Counter
- U6 - 74C221 Dual Monostable
- U8 - 74C926 4-Digit Counter w/Multiplexed 7-Segment Output Driver w/Carry Out
- U9 - 74C925 4-Digit Counter w/Multiplexed 7-Segment Output Driver
- U10 - CD4503B HEX Non-inverting TRI-STATE Buffer
- U12 - CD4081B Quad 2-Input AND Gate
- U13 - CD4071B Quad 2-Input OR Gate

Resistors: 1/4-Watt, 5%
- R1 - 3M
- R2-R4 - 22K
- R5-R18 - 100 Ohm
- R19 - 220 Ohm
- R20 - 2.2K
- R21, R22, R23 - 10K

Capacitors: Low Voltage Miniature Ceramic
- C1-C3 - 1000 pF
- C4 - .01 uF
- C5-C19 - .1 uF

Table 3.4.2. Parts List - Counter/Timer Circuit, Figure 3.4-2.

The frequency ranges are in three-switch selectable bands: 20 Hz - 20 KHz, 20 Hz - 50 KHz, and 20 Hz - 100 KHz. The frequency parameters may be varied to suit individual needs without changing the basic features of the construction. The ICL8038 offers an extended operating range of .001 Hz to 300 KHz.

The circuits employ ± 15 volt power to maintain a zero reference for the sine wave. Even so, I observed some offset and inserted a 100 mF capacitor. Five and fifteen volt square wave levels are referenced to ground. Note that all outputs are buffered. A level control is provided for the sine wave out. The sine output is a low impedance current driver, the LH0002. Also, control R4 is brought out to permit adjustment of the symmetry. This provides an improvement in distortion at frequencies below 100 Hz.

The Sawtooth Ramp

The sawtooth ramp circuitry is shown in the lower half of the schematic. The input is similar to that for the S-T-S with a notable difference: it is driven from a high resistance current source. This makes the response nonlinear in a direction to favor increased sensitivity as the ramp time is reduced.

The ramp time is switch selectable from a few milliseconds to about however far you care to stretch it out. The timing capacitance for the first two positions of switch S4 provide ranges of zero to 10 mS and zero to 100 mS. The third position connects to terminals for an external timing capacitor.

The timing capacitance is positioned in the negative feedback loop of integrator op amp U6B. In its operation the integrator must be periodically reset. The reset functions are carried out by the analog comparator, U7, dual monostable U8A and U8B, and analog switch U9A and U9B. The two switches are parallel connected which somewhat reduces the series resistance. Resistors R24 and R25 set the reference level at which the comparator output changes state, triggering the monostables to short the switches. This discharges the timing capacitors, the integrator output drops to zero and the cycle begins anew.

The sawtooth output is connected to two points: U12 (a Schmitt input inverter), and the top of ramp height control, R29. The inverted output connects to U5B, a CD4049 inverter. These provide a 15 volt to 5 volt conversion for the counter input. An LH0002 current driver is also used here.

The Counter/Timer

The generator incorporates its own six-digit counter for frequency and time measurement. *Figure 3.4.2* is the counter/timer circuit schematic. The inputs are switch selectable: S1 for the source, S-T-S or ramp; S2 for the

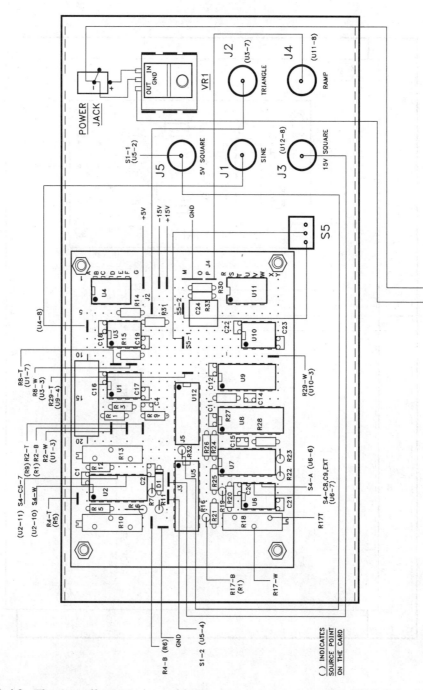

Figure 3.4.3. *The overall wiring assembly. Best approach is to use 24 AWG flat cable. Be reasonably generous with lengths. PART 1 OF IMAGE.*

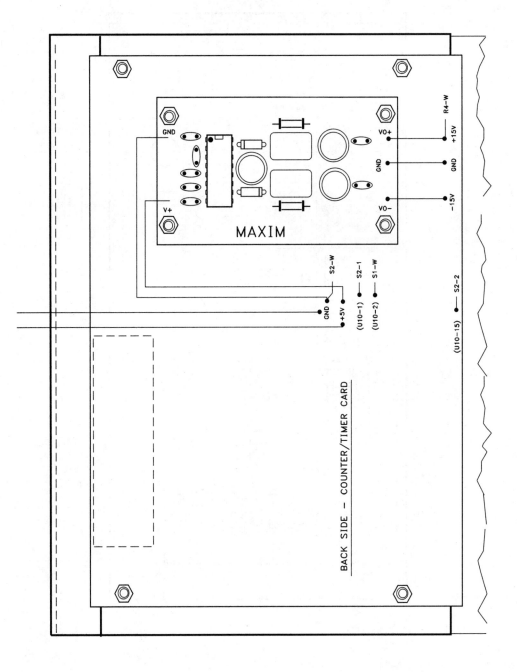

Figure 3.4.3. *PART 2 OF IMAGE.*

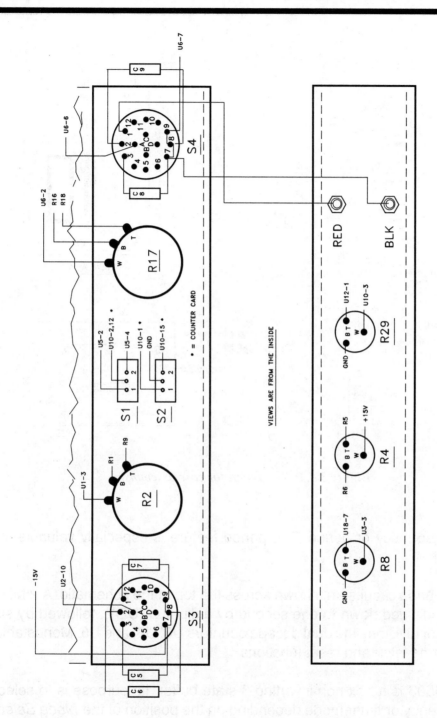

Figure 3.4.3. *PART 3 OF IMAGE.*

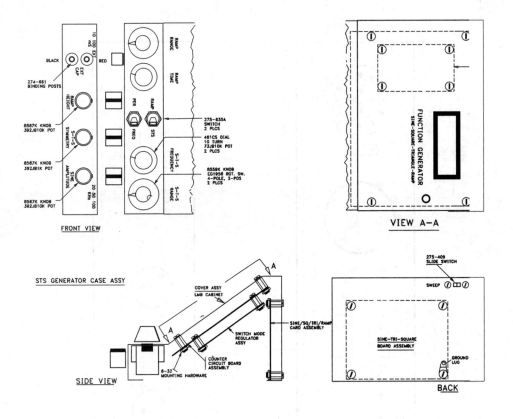

Figure 3.4.4. The generator case assembly.

mode: frequency or period. The period feature is especially valuable for ramp timing.

The time base circuits are shown across the top of the schematic. A 2 MHz crystal is divided down to one second by a flip-flop (U2A), followed by six decades of division: the dual decade counters U3, U4, and U5. Monostable U6 performs latch and reset functions.

The CD4503 is a hex non-inverting 3-state buffer. Its purpose is to select the frequency or timer mode depending on the position of the *Mode Select* switch S2.

Power Supply

The power source is a 9 VDC wall unit connecting to a rear panel jack. An internal 5-volt regulator and a $\pm$ 15 VDC converter provide the operating voltages. I used the Maxim[2] DC/DC converter for this. It resides on the backside of the counter/timer board assembly. Placement and wiring are shown in *Figure 3.4.3*.

Figure 3.4.3 shows the author's overall assembly and wiring diagram. It is highly advisable to fully connect all components and operate the generator prior to finalizing the case installation — 'cause it's a close fit!

* DS1-DS6 .300", Orange, 7-Segment Display, Common Cathode
* S1, S2 - Switch, Mini-Toggle, SPDT, Radio Shack 275-635B or Equal
* S3, S4 - Switch, Rotary, 3-Position, 4-Pole, Digi-Key EG-1958-ND
* S5 - Switch, Slide, Radio Shack 275-407 or Equal
* J1-J5 - Jack, BNC, UG-1094
* VR1 - LM7805CK Voltage Regulator, 5 VDC @ 1A
* Heatsink, TO-220, Jameco P/N 70771 or Equal
* Wall Transformer, 120 VAC to 9 VDC @ 1A, Jameco P/N 100888 or Equal
* J6 - Power Jack, Male, 11mm x 9mm x 14.5mm, Jameco P/N PP014 or Equal
* Knob, Round w/Dial, Digi-Key P/N 8559K-ND, 2 Reqd
* Concentric Scale Dial, Digi-Key P/N 461CS-ND, 2 Reqd
* Binding Post, 1 Each, Red and Black, Radio Shack 274-661A or Equal
* Cabinet, LMB MB 007-746, Sloping Front, 7 x 4 x 6-1/4" Gateway Electronics, 1-800-669-5810
* General Purpose Component PC Board, 2-13/16 x 3-11/16 Radio Shack P/N 276-158A
* General Purpose Component PC Board, 4-1/2 x 6-1/4 Radio Shack P/N 276-147A
* MAX743EVKIT - +5V to $\pm$ 15V Converter Kit, Available from Maxim Integrated Products, 120 San Gabriel Drive, Sunnyvale, CA 94086, (408) 737-7600

Table 3.4.3. Parts List - The Generator Case Assembly, Figure 3.4-4

2. *Maxim Integrated Products, Dual-Output, Switch-Mode Regulator, MAX743EVKIT Power-Supply Evaluation Kit.*

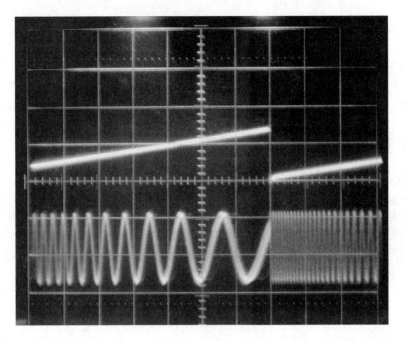

Photo 3.4.1. Sweep example using the ramp function. Upper trace: sweep sawtooth. Lower trace: swept frequency performance. Vertical scale: 10 V/div, both traces. Horizontal 1 mS/div. External trigger from the sawtooth.

Construction

I housed the electronics in an aluminum case with a sloping face for easy readability. The overall assembly is seen in *Figure 3.4.4*. All components are mounted on the front cover. This housing, which may not be for every-one, is compact for minimizing benchtop space. All controls are face mounted with the outputs on the back surface. A slide switch on the back connects the ramp output to the ICL8038 input for swept frequency use.

Application

An advantage of a versatile waveform generator is the ability to respond to "I wonder if" situations. The range of frequencies, the combination of sine and square wave outputs, with two voltage levels for the square wave make this unit adaptable to almost any experimenter's needs. The addition of the independent sawtooth ramp adds further to the versatility.

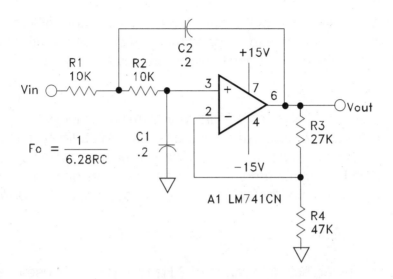

$$F_0 = \frac{1}{6.28RC}$$

Figure 3.4.5. *A low pass active filter for filtering distortion at low frequencies.*

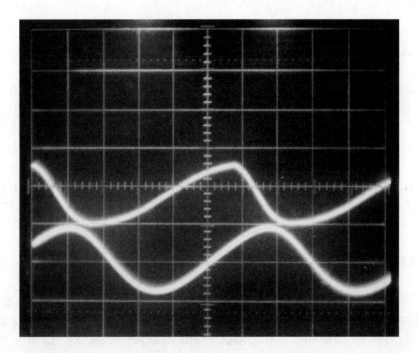

Photo 3.4.2. *Low frequency distortion reduction using a low-pass active filter, Figure 3.4.-5. Generator frequency 38 Hz. Vertical 10 V/div, both traces. Horizontal 5 mS/div.*

Photo 3.4.1 illustrates use of the ramp over a swept frequency range. The ramp can also be used as a long-term timer. In conjunction with an analog comparator and supportive logic it can be employed for level sensing and responsive control functions.

In Chapter 4, section 4.2.10, a versatile waveform burst module is described. This has application to response measurements with audio systems, active filters and hydrophones as examples. The low pass active filter circuit of *Figure 3.4.5* is helpful for reducing distortion at lower generator frequencies. This is shown in the "before" and "after" scope traces of *Photo 3.4.2*.

3.5. A FOUR-MODE/SIX-DIGIT COUNTER/TIMER[1]

A frequency counter, having a range of 2 Hz to 1 MHz, is a useful addition to one's kit of electronics. The ability to measure the period of cyclic waveforms in this range to a resolution of one microsecond is an added benefit. This counter/timer goes two steps further with modes for:

1. Totaling randomly occurring events.
2. Measuring the width of singly occurring, positive going pulses.

In many situations, the two timing modes may take the place of an oscilloscope.

When constructed using the cabinet and component boards described, a compact unit of a pleasing, professional appearance results. There is no real requirement to adhere to this construction; the circuit and wiring information provided may be applied to any packaging approach you may desire.

Circuit Functions

The internal functions of the counter-timer are shown in the block diagram of *Figure 3.5.1*. Details of the complete circuit are provided in *Figure 3.5.2*. The major functions are: an input buffer stage, selection of the counter mode, the one-second time base, counter latch and reset pulse generation,

---

1. *Portions are reprinted with permission from Popular Electronics Magazine, March 1996 issue. (C) Copyright © Gernsback Publications, Inc., 1995.*

6-DIGIT COUNTER FUNCTION DIAGRAM

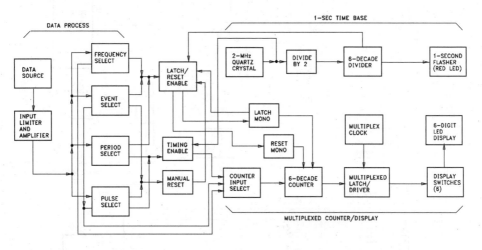

Figure 3.5.1. Counter/timer block diagram.

and the six-digit count and display. Function areas are denoted on the schematic. A 7555 timer circuit is included to provide a convenient internal means of verifying frequency count and period operation. Toggle switch S1 selects the test or normal input.

At the input, an LM339 comparator (U17) provides gain and waveform shaping. Diode D3 clips the negative going half of sinusoidal inputs. Diode D2 limits inputs to the five volt bus. Diode D4 provides a small input offset for comparator operation. U18, a CD4050, provides a buffer for the comparator output.

Rotary switch S2 selects the desired counter/timer mode. Note that the first two, *frequency* and *period*, respond to repetitive data while the remaining two, *event* and *pulse*, are event inputs.

The CD4503B is a tri-state, noninverting buffer. It is well suited as the means of channeling the input for the desired operating mode. The utility of the tri-state outputs is seen in devices U13 and U15. A high level on a disable input, pin 1 or 15, causes those gates on its control line to go into a high impedance state. Unused inputs are tied to ground, as noted on the schematic.

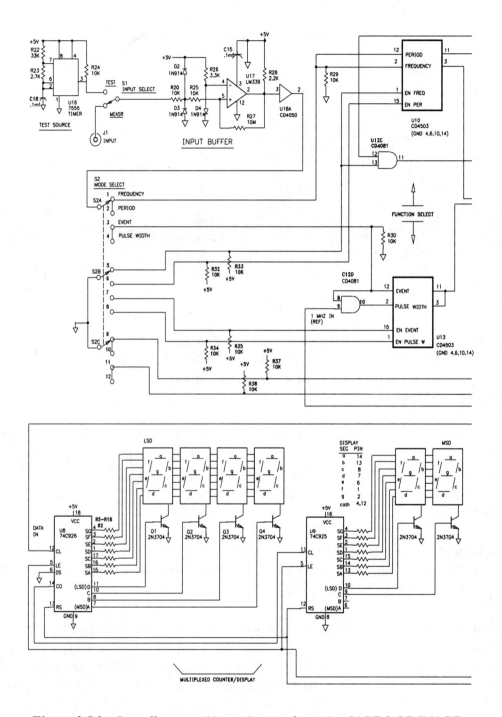

Figure 3.5.2. *Overall counter/timer circuit schematic. PART 1 OF IMAGE.*

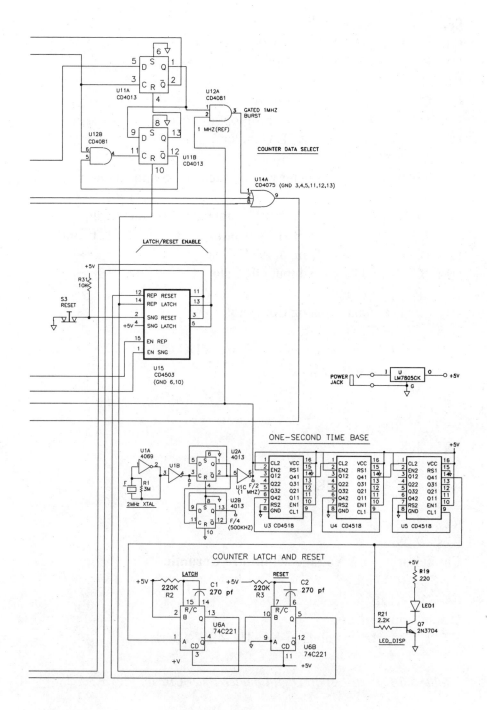

Figure 3.5.2. PART 2 OF IMAGE.

Semiconductors
- Q1-Q7 - Transistor, NPN, 2N3704 or Equal
- D1 - Red LED, MV5020 or Equal w/Mounting
- D2-D4 - IN914 or Equal
- U1 - CD4069B/74C04 Hex Inverter
- U2, U11 - CD4013B Dual D-Flip-Flop
- U3-U5 - CD4518B Dual BCD Counter
- U6, U7 - 74C221 Dual Monostable
- U8 - 74C926 4-Digit Counter w/Multiplexed 7-Segment Output Driver w/Carry Out
- U9 - 74C925 4-Digit Counter w/Multiplexed 7-Segment Output Driver
- U10, U13, U15 - CD4503B HEX Non-inverting TRI-STATE Buffer
- U12 - CD4081B Quad 2-Input AND Gate
- U14 - CD4075B Triple 3-Input OR Gate
- U16 - ICM7555 Timer
- U17 - LM339 Quad Analog Comparator
- U18 - CD4050B Hex Buffer
- VR1 - LM7805CK Voltage Regulator, 5 VDC @ 1A

Resistors: 1/4-watt, 5%
- R1 - 3M
- R2, R2-R4, R21 - 22K
- R5-R20 - 100
- R22 - 33K
- R23 - 2.7K
- R4, R5, R29-R31 - 10K
- R26 - 3.3K
- R27 - 10M
- R28 - 2.2K

Capacitors: Low Voltage Miniature Ceramic
- C1-C3 - 1000 pF
- C4-C19 - .1uF

Other Components
- XTAL - 2 MHz crystal, Digi-Key HC-49/VA Series

Table 3.5.1. Parts List - 6-Digit Counter/Timer Circuit, Figure 3.5-2.

The *frequency* and *period* measurements use the one-second time base and the sequential latch and reset pulses generated by it. These are switched out in the *event* and *pulse* modes. In these modes the counter latch input is held high. Reset is done manually with the normally closed push-button switch S3. The LED continues to flash at one second intervals however. This is an aid in pulse width measurements exceeding one second.

Let's consider the circuits associated with U10: AND gate U12 and the D-flip-flop U11. These two function in the *period* mode only. Their role is to limit the time measurement to a single cycle. The initial leading edge clocks the first stage of the flip-flop. The resulting high on Q1 enables U12A to gate one megahertz clock pulses to OR gate U14A for counting. The next input leading edge toggles the flip-flop second stage, locking the first stage in the reset mode, thereby shutting off the flow of clock pulses through U14. Gate U12B inhibits further pulses to U11B. Gate U12C inhibits the period operation when in the frequency mode.

IC U13 functions in the single input modes. For these inputs AND gate U12D pin 8 is held high to enable clock pulses to U13. The operation of U13 inhibits these except when in the *pulse* mode.

IC U15 differentiates the *repetitive* and *single input* latch and reset functions. The counter latch is held high when in the *event* or *pulse* modes; *reset* is performed manually by pressing switch S3. The counter displays counts to 999999. On the next input the display rolls over to 000000, accompanied by a flash of the LED. Long pulse widths can be measured by keeping track of the LED count. The widths of repeated pulses are summed.

The 74C925 (U9) and 74C926 (U8) are four-digit counters with internal output latching and multiplexed drivers for seven-segment displays. (A description is provided in Chapter 5.3.7) The 74C926 provides a carry out. The multiplexing operations are performed internally, providing a "rotating" pulse sequence. These are used to drive switching transistors, Q1 - Q6, for the display of each digit in its turn.

The one-second time base is derived from a 2 MHz crystal oscillator. Input to the one megahertz divider is derived from the first stage of U2; a 500

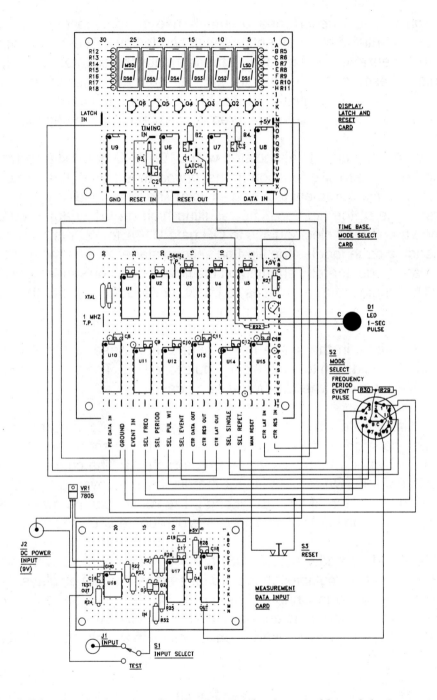

Figure 3.5.3. *Component parts layout of the author's assembly and the interconnection wiring.*

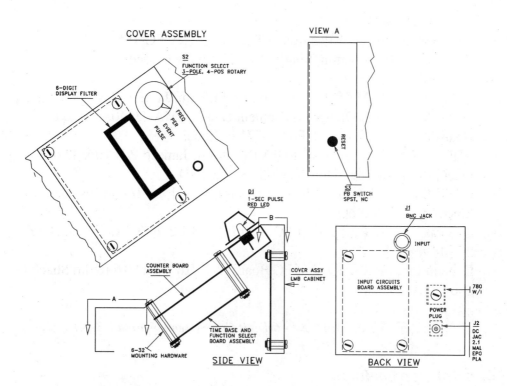

Figure 3.5.4. *The counter/timer housing assembly.*

KHz rate is also available from the second stage. This could be brought out if desired, buffered through an unused gate of the CD4050B, U18. It can be useful for verifying counter operation.

The one Megahertz is further divided by six stages of decade frequency division. Each CD4518 (U3, U4 and U5) provides two divide-by-ten stages. Latch and reset pulses are derived from the 74C221 monostable, U6.

Power Supply

The power supply is simplified by using a wall unit converting line voltage to filtered 9-volt DC power. An LM7805 provides a regulated 5-volt supply. Current drain is about 150 mA. All devices except the 74C925 and 74C926 will operate on up to 15 volts; these two ICs are restricted to 6 volts maximum. If desired, the regulator may be replaced with a six volt battery source for portability.

- S1 - Switch, Mini-Toggle, SPDT
- S2 - Switch, Rotary, 4-Position, 3-Pole, Digi-Key EG-1956 or Equal
- S3 - Switch, Push-button, Single Pole, Normally Closed
- J1 - Jack, BNC
- J2 - Power Jack, Male, 11mm x 9mm x 14.5mm, Jameco P/N PP014 or Equal
- DS1-DS6 .300 in., Orange, 7-Segment Display, MAN3640A or Equal
- Heatsink, TO-220, Jameco P/N 70771 or Equal
- Wall Transformer, 120 VAC to 9 VDC @ 1A, Jameco P/N 100888 or Equal
- Knob, Round w/Dial, Digi-Key P/N 8559K-ND
- Cabinet, LMB MB 007-446, Sloping Front, 4 x 4 x 6-1/4 inches, Gateway Electronics, 1-800-669-5810
- General Purpose Component PC Board, 1-7/8 x 2-29/32 Radio Shack P/N 276-149
- General Purpose Component PC Board, 2-13/16 x 3-11/16 Radio Shack P/N 276-158A, 2 each

Table 3.5.2. Parts List - 6-Digit Case Assembly, Figure 3.5-4.

Construction Features

The construction shown in **Figure 3.5.3** is proportioned between three circuit cards. The card layouts and interconnecting wiring are described in this figure. The component arrangement shown easily lends itself to point-to-point wiring.

The card arrangement was designed to accommodate a sloping face cabinet. The author's housing assembly is seen in **Figure 3.5.4.**

It is prudent to wire up the entire circuit on a jig and verify the operation in all four modes before putting it all together, especially if you use an assembly similar to the author's.

3.6. A MULTICHANNEL OSCILLOSCOPE SWITCH

Solving troubles with sequential circuits the likes of counters and decoders with your typical single- or dual-trace oscilloscope can be frustrating. How great to simultaneously display all four outputs of a counter IC, for instance.

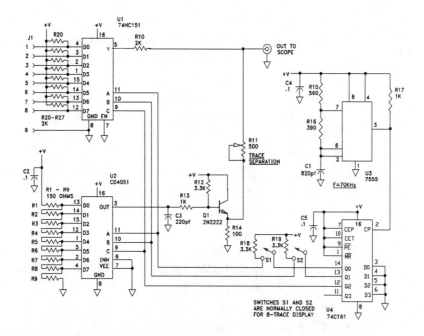

Figure 3.6.1. *Circuit schematic for the oscilloscope switch. Two, Four, or eight traces can be selected.*

Why, an erratic output would leap right off the screen for us; and for the decoded outputs as well, or the eight lines of a data bus or an analog-to-digital converter.

Nor are the advantages limited to troubleshooting; multiple traces are a great advantage in exploring the characteristics of these and similar circuits.

How It Works

The switch circuitry is described in *Figure 3.6.1*. The switching action is based on two CMOS eight channel multiplexers: a digital for the data to be displayed; an analog for the screen's vertical trace separation. These perform like 8-input, single-pole rotary switches. The "switch position" is set by the state of the three selection inputs, A, B, and C, the three least significant outputs of the 74C161 binary counter. A CMOS 7555 timer is the counter's clock source.

Semiconductors
- U1 74HC151 8-Channel Digital Multiplexer
- U2 CD4051 8-Channel Analog Multiplexer
- U3 7555 CMOS Timer
- U4 74C161 Synchronous Binary Counter
- Q1 2N2222 NPN Transistor, TO-92 Case

Resistors
- R1-R9 150 Ohm, 1/4W, 5%
- R10, R20-R27 2K, 1/4W, 5%
- R11 Trimpot 500 Ohm, Helitrim or Equal
- R12, R18, R19 3.3K, 1/4W, 5%
- R13, R17 1K, 1/4W, 5%
- R14 100 Ohm, 1/4W, 5%
- R15 560 Ohm, 1/4W, 5%
- R16 390 Ohm, 1/4W, 5%

Capacitors
- C1 820 pF, Ceramic
- C2, C4, C5 .1 mF, Ceramic
- C3 220 pF, Ceramic

Table 3.6.1. Parts List - Oscilloscope Switch Card.

On-screen trace separation is enabled by the transistor circuit. The voltage drops across resistors R1 - R9 provide an 8-step staircase at the CD4051 output. Waveforms are seen in *Photo 3.6.1*.

Pullup resistors R20 - R27 at the 74HC151 inputs ensure compatibility with most input sources. Connection to the points in the circuit of interest are made with mini test clips. I found 24-gauge, color-coded flat cable to be ideal for this. Connection to the switch case is via a 9-pin D-type subminiature connector. The connector can be omitted if straight-through wiring is preferred. The ninth input is utilized for connection to the circuit ground.

Switch contacts S1 and S2 are seen in series with the two least significant counter outputs, Q0 and Q1, respectively. (***Figure 3.6.1***) Normally these are closed for full 8-trace multiplexing. At times the screen is easier to read

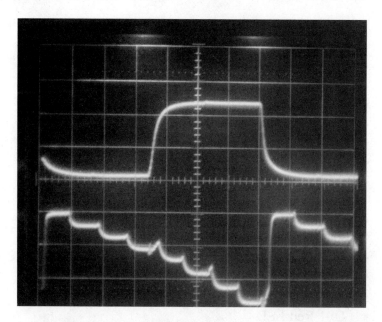

Photo 3.6.1. *Upper trace: one clock cycle, output Q2 at U4 (74C161), pin 12. Lower trace: The staircase seen at pin 3 of U2 (CD4051). Time/div. - 1.5 uS for a roughly 10 uS switch cycle time.*

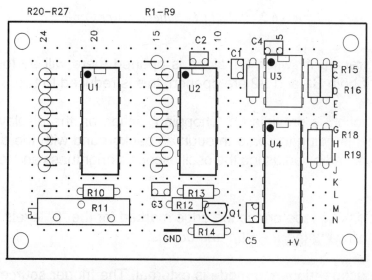

BASED ON RADIO SHACK BOARD 276-149A

Figure 3.6.2. *The circuit board parts layout for the switch components. The board wiring is point-to-point, but a printed circuit card could be easily made.*

Resistors
- R1 150K, 1/4W, 5%
- R2A Trimpot 200 Ohm, Helitrim or Equal
- R2B 470K, 1/4W, 5%
- R3 3.0M, 1/4W, 5%
- R4 220 Ohm, 1/4W, 5%
- R5 270K, 1/4W, 5%
- R6 10K, 1/4W, 5%
- R7 1K, 1/4W, 5%

Capacitors
- C1 1.5 mF, 15V, Tantalum

Semiconductors
- U1 ICL8211 Programmable Voltage Reference, Harris
- U2 7555 CMOS Timer
- Q1 2N2222 NPN Transistor, TO-92 Case
- DS1 Red LED
- DS2 Amber LED

Other
- S1 Toggle Switch, SPST, Radio Shack 275-634B or Eq.

Table 3.6.2. Parts List - Battery Power Card..

with fewer traces. If switch S1 alone is opened there will be four traces displayed. If both S1 and S2 are opened, this is reduced to two.

The clock action is providing a "chopping" action on the displayed data. Some of this is going to come through as background with the data. This can be minimized by adjusting the oscilloscope's brightness and focus controls.

A strip of masking tape on the left edge marked for the "0" levels can be a real aid in following the *0 - 1 up and downs.*

In use, the scope triggering mode is external. The trigger source is in the circuit you are working with. Move the trigger source around to bring out a most informative display.

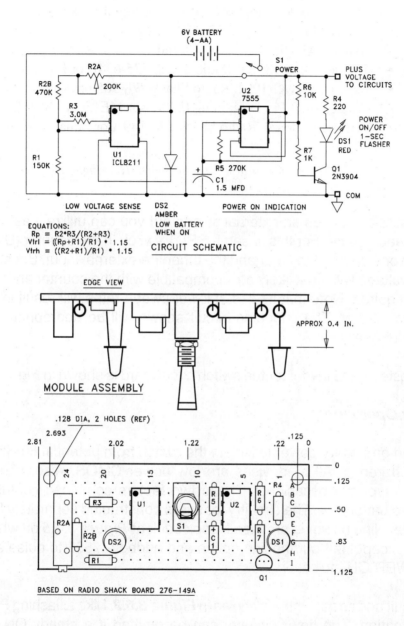

EQUATIONS:
$$R_p = R2 \cdot R3/(R2+R3)$$
$$V_{trl} = ((R_p+R1)/R1) \cdot 1.15$$
$$V_{trh} = ((R2+R1)/R1) \cdot 1.15$$

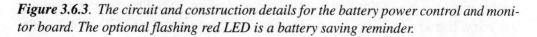

Figure 3.6.3. *The circuit and construction details for the battery power control and monitor board. The optional flashing red LED is a battery saving reminder.*

- Case, Plastic, Radio Shack 270-223 or Equal
- Battery Holder, Mouser 12BH447 or Equal
- Battery Snap, Mouser 12BC010 or Equal
- Battery, Alkaline Size AA, 4 Reqd.
- Toggle Switch, SPST, Radio Shack 275-634B or Eq., 2 Reqd.
- Jack, BNC, UG-1094, Radio Shack 278-105 or Equal
- Connector, 9-Pin D-Submin, F, Mouser ME156-1309 or Equal
- Perforated Board, Radio Shack 276-149A or Equal

Table 3.6.3. Parts list - the Switch Housing.

If your scope provides an external sweep out you can utilize it as an alternate sweep source. For this, insert a SPDT switch in the counter (U4) input line with one source the timer and the other the external input. Because the sweep voltage will most likely be incompatible with the counter an interface will be required. Note that its use at lower sweep rates will result in significant trace flicker. The chopping circuit shown will be appropriate for the majority of needs.

A suggested parts layout for the switch circuit card is shown in **Figure 3.6.2**.

Battery Operation

I decided on battery power to isolate the switch from potential 60 Hz power line interference. All IC devices are low power CMOS. Also, use of the switch is typically occasional. These two factors combined contribute to extended battery life. The circuitry for this is constructed separately for convenience if line power is preferred. Look back at **Figure 3.1.5** on which it is based, except I left out the line option and added a 7555 to pulse an LED for POWER ON indication.

The circuit and its assembly is shown in **Figure 3.6.3**. I like a flashing POWER ON indication. The timer circuitry can be omitted if a steady ON light is preferable. Some indication is recommended as a low battery reminder. I adjusted the low voltage set point to switch on the amber LED at 4.3 volts.

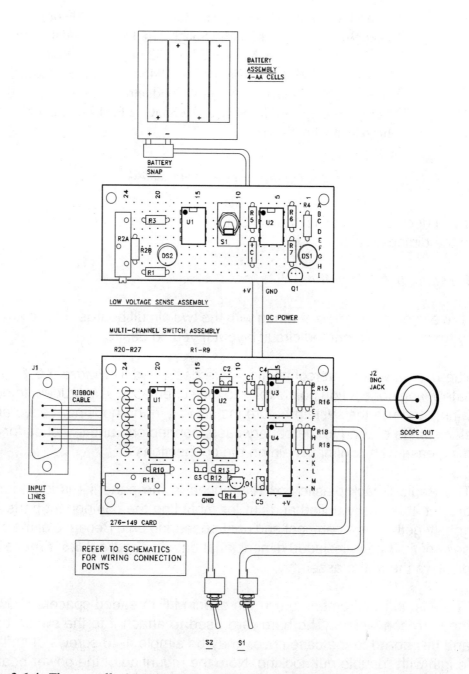

Figure 3.6.4. *The overall wiring interconnections. It's a good idea to wire it all and check the performance before installing in the case. Leave adequate slack in the wiring. 24 gauge color coded ribbon cable is easiest to work with.*

- Connector, 9-Pin D-Submin, M, Mouser ME156-1209 or Equal
- Hood, 9-Pin D-Submin, Radio Shack 276-1513 or Equal
- Jack Screw, Keystone, Mouser 534-7232 or Equal, 2 Reqd.
- Spacer, 5/8", Keystone, Mouser 534-1945 or Equal, 6 Reqd.
- Ribbon Cable, 24 Gauge, 24" of 9 Conductors
- Micro Test Clips, Radio Shack 270-355C, 2 Packs of 4 Each
- Micro Test Clip, Single, 1

Table 3.6.4. Parts List - the Flat Cable Assembly.

If you decide to stick with the construction scheme it is essential to adhere to the dimensions shown.

Bringing It All Together

I have used point-to-point wiring with the two circuit boards. These are easily translated to a printed circuit layout if you so desire.

The overall wiring interconnections are described in **Figure 3.6.4**. To locate wiring connection points refer to the schematics. It is prudent to do the wiring and run the system before installing it in the housing. Leave ample slack in the cabling to the battery case, connectors, and switches for handling ease with minimal strain on the connections.

The Radio Shack plastic box referenced in the parts list is ideal for this project. It is easily drilled and cut for mounting the component parts. It all goes together as a compact and neat appearing unit. Because of the closeness of fits I have included dimensional data in **Figure 3.6.5**. **Figure 3.6.6** displays the entire assembly.

There is no requirement to use the exact M/F threaded spacers shown — they are convenient. The hardware used to attach it to the switch board, and this board to the case cover, may be simple 4-40 screws of sufficient length with "double nut" locking. Note the mounting of the power board by its toggle switch to the case cover. The interior depth of the case is 1.75 inches. The assembly as shown is 1.5 inches deep. It looks really great when it's all together.

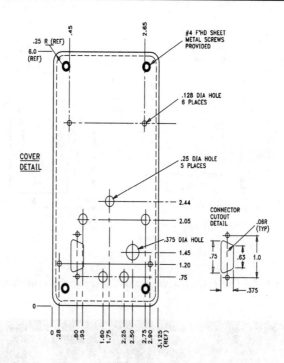

Figure 3.6.5. *Dimensional details of the case cover.*

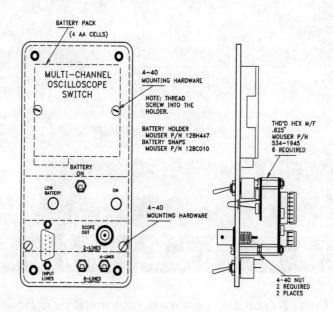

Figure 3.6.6. *Case cover components assembly. All parts mount on the cover for convenient access. Take careful note of how the two cards are supported.*

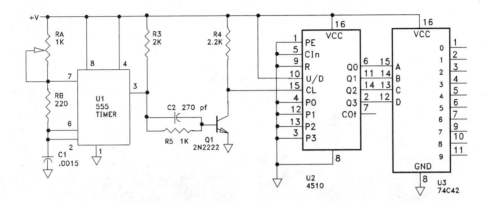

Figure 3.6.7. *The circuit schematic used for the counter/decoder display.*

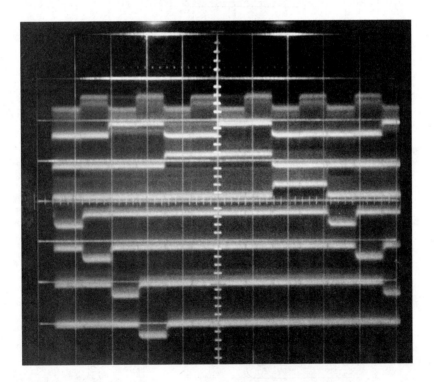

Photo 3.6.2. *Traces 1-4 (top down), are counter outputs (Refer to Figure 3.6-7). Traces 5-8 are the initial four outputs of decoder 74C42. Scope triggering from decoder output 0, (74C42, pin 1). Vertical sensitivity - 50 mV/div. Time/div - 10 uS.*

Application Examples

A valuable use of the switch is the exploration of counter and decoder circuits. A simple circuit example is shown in **Figure 3.6.7**. The four traces for the counter outputs and traces for the first four decoder outputs are shown in **Photo 3.6.2**.

Experiments with analog-to-digital and digital-to-analog circuits are a valuable use of the switch. A test circuit for an 8-bit conversion is shown in **Figure 3.6.8**. (The *Precision Constant Voltage/Current Source*, section 3.3, is ideal for the *Adjustable Voltage Source*.) The bench setup I used for the observation is shown in **Figure 3.6.9**. The screen traces of *Photo 3.6-3* show two conditions: the first with all inputs at ground as a reference and the second with an analog input voltage.

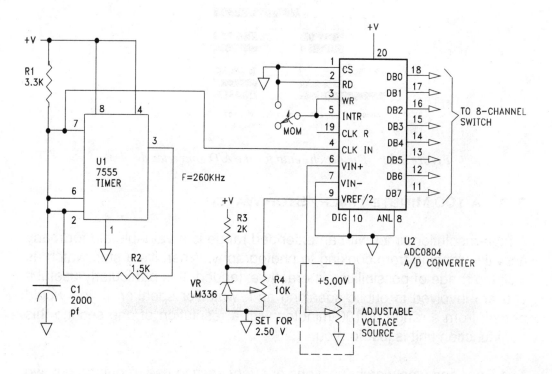

Figure 3.6.8. *The circuit schematic used for the A/D converter display.*

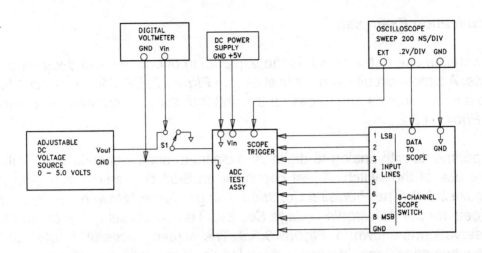

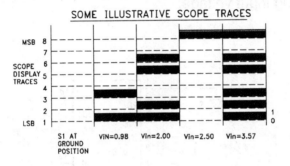

Figure 3.6.9. The bench setup for the A/D converter display.

3.7. A 100 MINUTE TIMER/STOP WATCH

A high-resolution timer with an extended range is a valuable aid for many activities ranging from cooking to photography. When it's a stop watch as well the range of possibilities is greatly extended. It is particularly useful if you are involved in public speaking as a writer or speaker. If you would benefit from a range of 100 minutes with a resolution of one second this dual function unit is just for you.

The timer and stop watch functions are very easy to use, requiring but two switches for its control. One, a momentary push-button, resets the counter and initiates the timing sequence. The second is for holding the count at the current value — enabling the stop watch feature.

The timer is preset via four panel mounted thumbwheel style switches with BCD coding. Each switch section is provided with two push-buttons for incrementing/decrementing the setting. The range of each section is 0 - 9 in increments of one count. Two switches yield a range of 0 - 99 minutes; the remaining two of 0 - 59 seconds.

The unit is normally battery operated for portability but may be powered from the line by connecting a wall unit having a 9-volt DC output.

Circuit Features

The circuit diagram for the timer is shown in *Figure 3.7.1*. A first impression may be that this is a complex circuit. That is not really so; basically it is a combination of readily understood functions.

The focus of the circuit is the selection of the timer delay and the display of the elapsed time. These functions are supported by a clock source, timer initialization, count decoding, and display multiplexing requirements. The various function blocks are identified on the schematic.

The elapsed time is displayed on a 4-digit, multiplexed, 7-segment LED display module. All integrated circuits are CMOS for battery conservation. A 7555 timer IC, U20, and a CD4017 decade divider/decoder, U21, provide the multiplexer clocking. The frequency for the values shown is about 220 Hz.

The clock oscillator is a 7555 timer IC, U1, operating at 60 Hz. The schematic shows two values of timing capacitance. The smaller speeds up the clock by a factor of ten, a real time saver when getting the unit up and running. When it is fully operational the second capacitor is connected in parallel. At this time trimmer R2B is adjusted for 60 Hz.

A CD4040 12-bit binary counter, U2, and a CD4082 Dual 4-input AND gate, U3, decode the incoming clock train for a one-minute and one-second output. The one-second output provides the clock source for the CD4518 dual BCD counter, U7. The minute output resets U2 via a CD4071 two-input OR

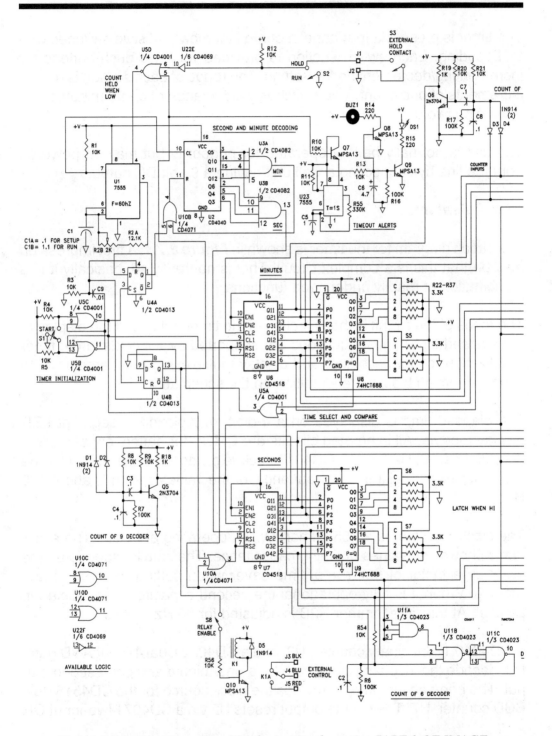

Figure 3.7.1. *The Timer/Stop Watch circuit schematic. PART 1 OF IMAGE.*

100 MINUTE TIMER

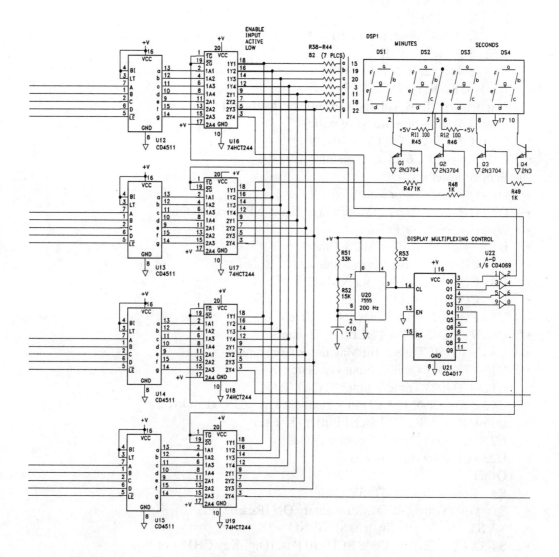

Figure 3.7.1. *PART 2 OF IMAGE.*

Resistors: 1/4W, 5% Unless Otherwise Noted

- R1, R3, R4, R5, R8, R9, R10, R11, R12, R13, R20, R21, R54, R56 - 10K
- R2A - 12.1K, 1%
- R2B - 2K Cermet Trimpot
- R6, R7, R16, R17 - 100K
- R14, R15 - 220
- R18, R19, R47, R48, R49, R50 - 1K
- R22 - R37, R53 - 3.3K
- R38 - R44 - 82, R45, R46 - 100
- R51 - 33K, R52 - 15K, R55 - 330K

Capacitors: Low Voltage Ceramic or Tantalum
(Noted With + Sign on Schematic)

- C1A, C2, C3, C4, C7, C8, C10 - .1
- C1B, C5 - 1.0 C6 - 4.7 (Tantalum) C9 - .01

Semiconductors

- Q1 - Q6 - 2N3704 or Equal
- Q7 - Q10 - MAPSA13 (Motorola or Equal NPN Darlington)
- D1-D5 - 1N914 or Equal
- U1, U20, U23 - 7555 CMOS Timer
- U2 - CD4040 12-Bit Binary Counter
- U3 - CD4082 Dual 4-Input AND Gate
- U4 - CD4013 Dual D Flip-Flop
- U5 - CD4001 Quad 2-Input NOR Gate
- U6,U7 - CD4518 Dual BCD Up Counter
- U8,U9 - 74HCT688 8-Bit Magnitude Comparator
- U10 - CD4071 Quad 2-Input OR Gate
- U11 - CD4023 Triple 3-Input NAND Gate
- U12-U15 - CD4511 BCD to 7-Segment Latch/Decoder Driver
- U16-U19 - 74HCT244 Octal Buffer (3-state)
- U21 - CD4017 Decade Counter/Divider
- U22 - CD4069 Hex Inverter

Other

- K1 - Reed Relay, SPDT, 5V Coil
- S1 - Push Switch, SPDT, Momentary ON (Reset/Start)
- S2, S3, S8 - Toggle Switch, SPST, NO
- S4-S7 - TW Switch, Code BCD, 10 Pos. (Digi-Key CH181 or Eq)
- DSP1 - 4-digit LED Display, Com. Cath. (Digi-Key P454 or Eq)
- J1 - J5 - Insulated Binding Posts w/Banana Jacks (Radio Shack 274-661a or Equal)
- DS1 - LED, Red Super Bright, (Radio Shack 276-088 or Equal)
- BUZ1 - 3.6KAZ Piezo Buzzer (Radio Shack 273-060 or Equal)

Table 3.7.1. Parts List - Timer/Stop Watch circuit schematic.

gate, U10B. The other gate input is for front panel startup initialization and reset.

Since 90-second minutes are not allowed (for obvious reasons), the first stage of U7 must be reset at the count of 60. The gates of CD4023, a triple 3-input NAND gate, U11, provide the decoding. The delay network, R6, R54, and C2 was found necessary to ensure reliable comparison and counting. The decoder output resets both stages of U7 via another OR gate, U10A. It also clocks the first stage of the minute counter, U6, also a CD4518.

To my surprise, I found it necessary to provide "count of nine" decoding on the four counter outputs, again to ensure reliable recognition. For these I employed a simple 2-input diode-transistor OR circuit with a brief time delay.

The thumbwheel switch settings are monitored by two 74HCT688 8-bit magnitude comparators, U8 and U9. When the eight counter outputs correspond with the switch settings, comparator pin 19 goes LOW. These LOW states are input to a CD4001 2-input NOR gate, U5A. When both inputs go low, U5A clocks the CD4013 dual D flip-flop, U4B. Both sides of U4 are initially reset by R3 and C9 to ensure their being in the required state. The U4B Q output is input to U5D. The output of this NOR gate goes low with the input HIGH. The LOW state inhibits the operation of U1, holding the clock count at its present value. The count is displayed until reset externally. In the timer mode, this is the value preset on the four switches.

The CD4001 latch circuit, U5B and U5C, provides for initial zeroing and resetting of the counters. Switch S1 is a momentary ON SPDT push button. Pressing S1 at any time resets the counters. With hindsight, the "A" side of U4 is not necessary and could be omitted.

A second switch, S2, a SPDT toggle, provides a *RUN/HOLD* feature. The *RUN* position is normal for this switch; when positioned to *HOLD* the clock is stopped with the count preserved on the display. It is this feature that provides the stop watch function. Just ensure the thumbwheels are set to a value greater than any anticipated time. Front panel binding posts are provided for remote execution of this function. Note also that five volt reed

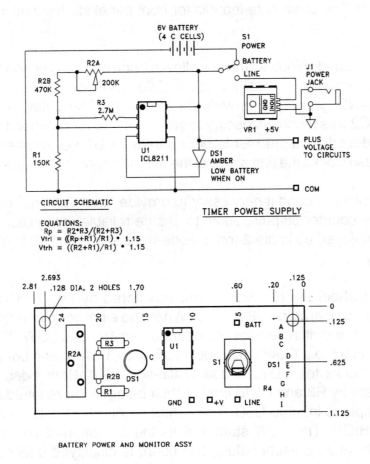

CIRCUIT SCHEMATIC

TIMER POWER SUPPLY

EQUATIONS:
$$Rp = R2 \cdot R3/(R2+R3)$$
$$Vtrl = ((Rp+R1)/R1) \cdot 1.15$$
$$Vtrh = ((R2+R1)/R1) \cdot 1.15$$

BATTERY POWER AND MONITOR ASSY

Figure 3.7.2. *The power supply and low voltage detection and display circuit schematic and assembly. Note the similarity to Figure 4.2-14.*

relay K1 provides for external control on time-out. The relay switch, Darlington transistor Q10, is preceded by an enabling SPST toggle, S8, for battery conservation.

The multiplexed display presented an initial challenge in requiring two independent sources of differing count limits. The solution is in the CD4511 BCD to 7-segment latch/decoder drivers, U12-U15, in conjunction with 74HCT244 3-state octal buffers, U16-U19. The decoded outputs of U21 are inverted by the CD4069s, U22A-U22D, and directed to the buffer en-

Resistors: 1/4W, 5% Unless Noted
- R1 - 150K
- R2A - 200K Cermet Trimpot
- R2B - 470K
- R3 - 2.7M

Semiconductors
- DS1 - LED (Yellow)
- U1 - ICL8211 Programmable Voltage Detector
- VR1 - 7805 +5V Regulator

Other
- S1 - Toggle Switch, SPDT
- J1 - Power Jack Compatible w/9V AC Adapter (Jameco AD 901A or Equal)
- Batteries - 4 Size C Cells
- Battery Holder (Radio Shack 270-390A or Equal)
- Small Heatsink for VR1

Table 3.7.2. Parts List - Power/Monitor circuit schematic, Figure 3.7-2.

ables, pins 1 and 19. The buffers sequence transistor switches Q1-Q4 and select the seven data inputs for display. The display is a single 4-digit common cathode module. The two leftmost digits show the elapsed minutes. A colon separates the two seconds digits on the right.

In the timer mode time-out results in a flashing high-intensity red LED and a pulsating buzzer. (From experience, you may want to include a switch in series with the buzzer.) For this, another 7555 timer (U23) is enabled by Q7, a darlington transistor. The transistor base is pulled HIGH by the Q output of U4B.

Circuit Power

Normal operation is maintained by four C-size alkaline 1.5 volt cells. This enhances the portability. However, frequent longtime timing usage may deplete the batteries unnecessarily. Line operation is provided by the connection of a nine volt wall unit adapter and an internal five volt regulator, VR1. The circuitry is shown in *Figure 3.7.2*. Also provided is the optional

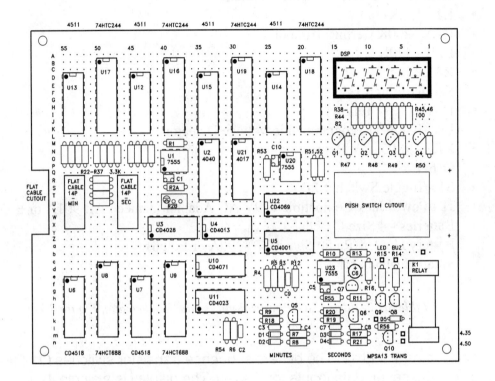

Figure 3.7.3. *The author's component board layout. The board shown is the Radio Shack 276-147a.*

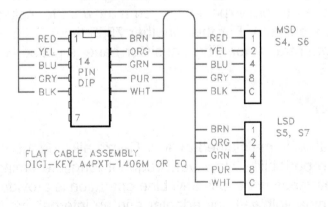

Figure 3.7.4. *The thumbwheel switch flat cable wiring diagram.*

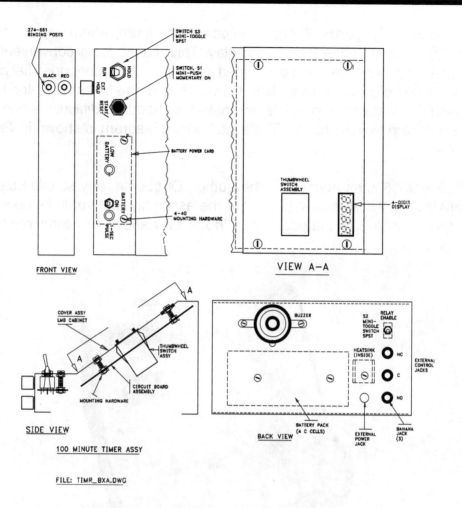

Figure 3.7.5. *The author's housing assembly.*

low battery voltage monitor and indicator circuit employing the ICL8211 programmable voltage detector described previously.

Construction

I employed perforated component boards with point-to-point wiring. The primary board layout used is shown in ***Figure 3.7.3***. A smaller, separate, assembly for the power control and monitor parts is shown in ***Figure 3.7.2***.

The layout of *Figure 3.7.3* accommodates the thumbwheel switch assembly in a cutout in line with the display. This makes for a convenient and pleasing appearance of the completed assembly. Two color-coded flat cable/14-pin DIP plug assemblies link the switches to the board circuitry. Hardwiring the switches is possible but makes for handling difficulties and possible hazard to the circuitry. The switch wiring diagram is shown in *Figure 3.7.4*.

Figure 3.7.5 is as housed by the author. Of course, any suitable case is satisfactory. As always, it pays wire the assembly and verify its operation before wrapping up the package. *Photo 3.7.1* shows the complete timer/stop watch.

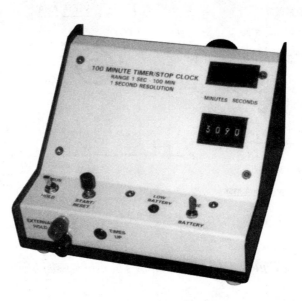

Photo 3.7.1. The assembled 100-minute timer/stop watch.

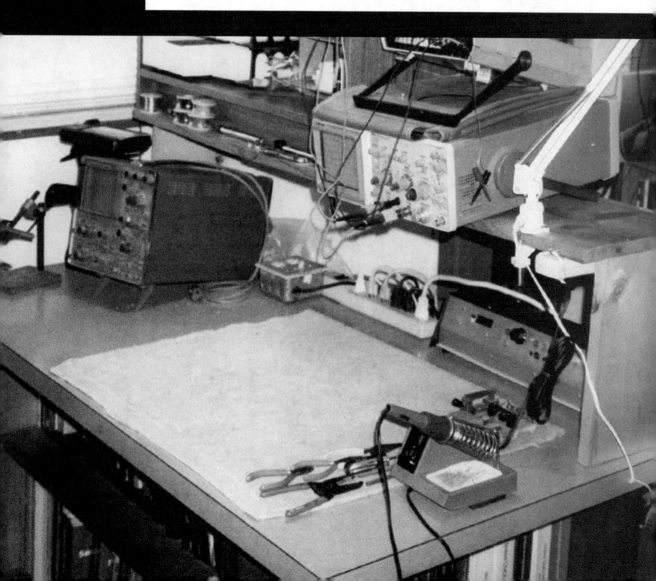

CHAPTER 4

BREADBOARDS &
PROTOTYPE MODULES

CHAPTER 4
BREADBOARDS
& PROTOTYPE
MODULES

The only way possible for me to write this book is to believe that you, the reader, feel much as I do about the joy of experimenting with electronic projects. A big portion of that enjoyment comes from plugging real parts into a real circuit strip, then connecting it all together — in part from scribbles on a scrap of paper and in part from ideas still riding around in the formation stage. Then comes that big moment when we turn on the power and... whoopee!

Sometimes yes, often no.

Very often the breadboard leads to thoughts of a better way to do the job. So, with all this in mind, let's move on to the whys and wherefores.

4.1. SOME WHYS AND MEANS

In the beginning, electronically speaking, there was the vacuum tube. Though the tube was large, it had very few I/O ports, to which connections were made via pin contacts on a plug-in socket. The tube operated with high voltages, which in general compelled connecting parts — capacitors and resistors and coils and such — to be large. There was a fair amount of heat to be dissipated. Experimenters took to driving nails into a piece of wood and soldering parts to these as connection points.

Which is why the term *breadboard* is not so silly at all.

As for the question "Why breadboard?", there is one good answer, which is, "Just to be sure our circuit performs as we expect." There is a corollary: very often, the breadboard points the way to a better solution.

4.1.1. Circuit Development

Several software programs for circuit development are available. I have a couple of these and they can be very useful. A real advantage is the savings in parts purchases. A disadvantage is the need for an extensive parts library which must be updated to remain current. For myself, as a hands-on person, I prefer working with the real parts.

Those of us working with circuits are fortunate in our choice of breadboarding materials. With the growing popularity of the transistor in the early sixties there arose a need for more efficient breadboarding. Many of us in those days simply solder-tacked the parts lead-to-lead, resulting at times in balloon-like masses of circuitry. Various items purporting to enhance our circuit developments popped up on the market from time to time; most with a well deserved brevity.

Then, in the early 1970s, as I recollect, the granddaddy of the solderless strip made its debut. With the rapid expansion of the integrated circuit in both the digital and analog arena this was a welcome development.

The *Experimenter 350*[1] socket by Radio Shack, shown in **Figure 4.1.1**, is an example of what is currently available. The item shown is convenient when putting together a small circuit, such as a 7555 clock. The socket is provided with end slots that enable linking a series of these end-to-end. Radio Shack also offers larger sizes.

In general, the sockets are best used with AWG 22 solid wire. Most part leads will fit. Economically priced kits of pre-cut, preformed, and color-coded wires are available. I augment these by making my own using color coding to separate the various wire lengths.

1. *A product of Radio Shack, a Division of Tandy Corporation, Fort Worth, Texas.*

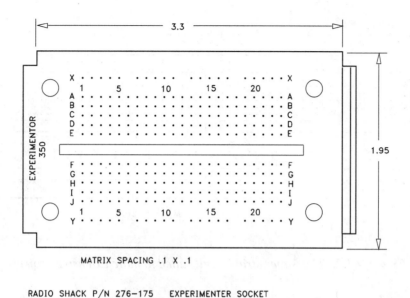

MATRIX SPACING .1 X .1

RADIO SHACK P/N 276-175   EXPERIMENTER SOCKET

Figure 4.1.1. *The Experimenter 350 socket by Radio Shack that is convenient when putting together a small circuit*

The Super Strip[2], shown in *Figure 4.1.2*, is suited to larger projects. Few precautions are needed for most digital circuits. However with sensitive analog circuits care must be taken in how the parts and connecting wiring are positioned.

While the figures illustrate but two variations, many other sizes are available. For myself, I mount these on a supporting surface, primarily to provide a backing as over time I have found some contact points have pushed down through, loosening the back covering, with a resultant unreliable connection.

A variety of development systems are also available with DC power, along with other supportive circuitry. Just about any need can be met. Two well established companies in this area are the *3M Company Electronics Products Division,* and the *Vector Electronic Company.*

2. *Super Strips and Circuit Strips are 3M Company Electronics Products Division universal breadboarding element with solderless plug-in tie points.*

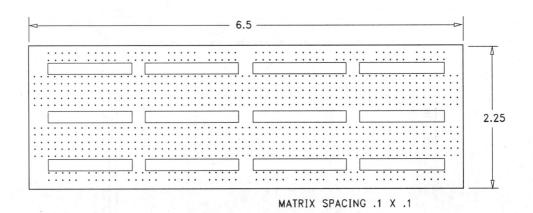

MATRIX SPACING .1 X .1

Figure 4.1.2. *This super strip is well suited to more extensive projects.*

4.1.2. Analog and Digital Examples

I have found breadboarding a circuit idea to be the really fun part of circuit development — also, at times, the most infuriating. Following the progress of an idea that may have kept me awake until two in the morning to watching its performance on the oscilloscope, is my version of climbing Mount Everest.

Unfortunately, breadboarding, as with mountain climbing, has its risks and setbacks. I have smoked my share or more of ICs with reversed power supply leads. I have stared at the circuit and its schematic for what has seemed like endless hours trying to understand why such an elementary collection of parts and wires is not fulfilling my expectations.

The culprit is usually one of three possibilities: it was a lousy idea; the connection to point B is at point Z; the connection linking points D and E isn't. Sometimes it means I had disconnected the power leads while making a change and they are still disconnected. Or the power supply never got turned on in the first place. Or was it the oscilloscope?

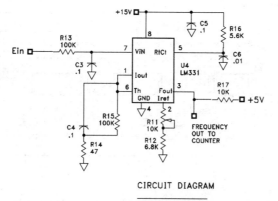

CIRCUIT DIAGRAM

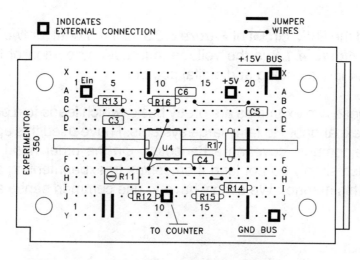

RADIO SHACK P/N 276-175   EXPERIMENTER SOCKET

Figure 4.1.3. *This voltage-to-frequency set up on the Experimenter 350 illustrates the ease with which even an analog circuit can be readily assembled.*

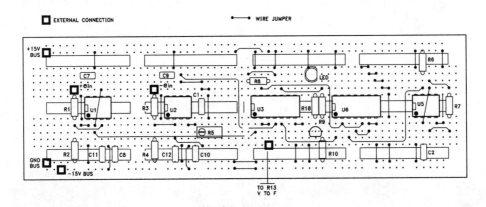

Figure 4.1.4. This mix of analog and digital circuitry is easily breadboarded. It pays to spread the circuit out as shown to allow for changes that often come to mind as the circuit development progresses.

I have used the DVM circuit of *Figure 3.2.3* as the model for two illustrative examples. *Figure 4.1.3* is the voltage-to-frequency portion of the circuit. *Figure 4.1.4* is the analog and digital portions.

Granted, these two examples are pretty basic, which leads to a point: these basics are expandable to fairly large systems. I have used this approach to circuit developments covering a benchtop. On the other hand, there are circuits which simply do not lend themselves to experimenting and development in this manner. We simply have to use common sense and be selective.

4.2. MODULES

4.2.1. An Adjustable Transistor DC Load

We've just constructed a power supply, say, perhaps one from Chapter 3. And we'd like to run a quick test to see what it can do by way of load current. This little module saves us countless hours of searching through our junk boxes for a power resistor(s) of just the right size and the required wattage.

In one evening's time you can assemble this variable load; adjustable over a range of zero to about one amp. *Figure 4.2.1* shows it all. The 2N3767 is able to withstand up to 80 volts with a current to four amperes. Rated dissipation, when properly heat sinked, is four watts. Of course, the design is not limited to the 2N3767; any suitable power transistor will do. *Photo 4.2.1* shows the assembled adjustable transistor DC load.

To use, connect a multimeter with a suitable current range to the *METER* terminals, the load source to be tested to the *SOURCE* terminals. Be sure to observe polarities. It is smart to have the current setting near zero initially. When all is ready flip the power toggle to ON. Now adjust the current to the desired level.

Although capable of DC current loading only, transformer secondary tests can be run by inserting a half- or full-wave rectifier in the circuit — also great for running battery performance tests under varying load conditions, or the output of the power amplifier you just did a fix on.

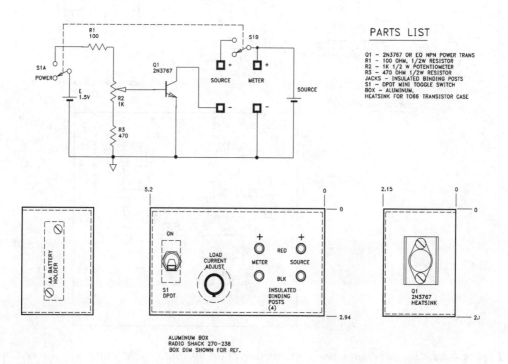

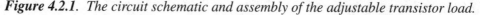

Figure 4.2.1. The circuit schematic and assembly of the adjustable transistor load.

Photo 4.2.1. *The assembled adjustable transistor DC load.*

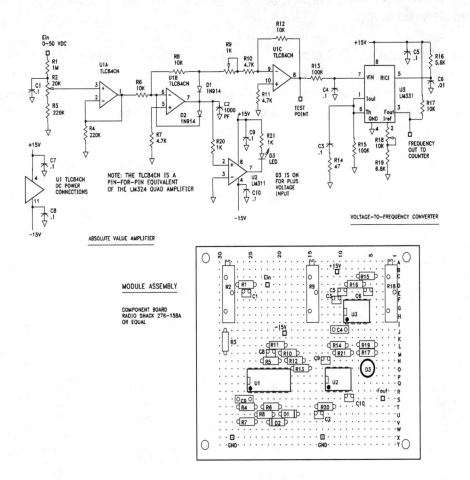

Figure 4.2.2. *The schematic and construction of an absolute value amplifier followed by a voltage-to-frequency converter.*

Nor is the unit limited to load testing. As you can see in Chapter 5.4.1, it is also a handy constant current source. The possibilities go on and on.

4.2.2. Absolute Value Amplifier with a Digital Output

This is a versatile module of many uses whenever simple conversion of a DC voltage to digital is needed. The circuit, shown in **Figure 4.2.2**, is an *absolute value amplifier* followed by a *voltage-to-frequency converter.* When used with a suitable display, such as that of section 4.2.5, this module performs as a simple DC digital voltmeter.

Chapter 3, section 2, provides a description of the amplifier's functions. **Figure 4.2.2** is a repeat of the circuitry with a useful component layout scheme included.

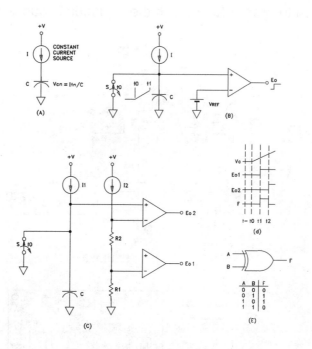

Figure 4.2.3. (a) Linear capacitor charging from a constant current source. (b) Use of a comparator to sense the one volt charge on the capacitor. (c) Two comparators define the incremental voltage. (d) Resultant waveforms. (e) The exclusive-OR logic gate truth-table.

4.2.3. Capacitance Measurement

In the mid-1970s, I acquired a *KIM-1* single board computer. I found I could use it for a variety of fascinating projects. An early one was the measurement of capacitance by precision charging of the unknown capacitor for one volt and converting the charging time to capacitance by the relationship:

$$Vcn = (I/C) \bullet tn$$

— where I is the charging current, C is the capacitance, and tn is the time to charge to one volt. From this we find:

$$C = (I/Vcn) \bullet tn$$

If we force Vcn to precisely one volt the equation reduces to:

$$C = I \bullet tn$$

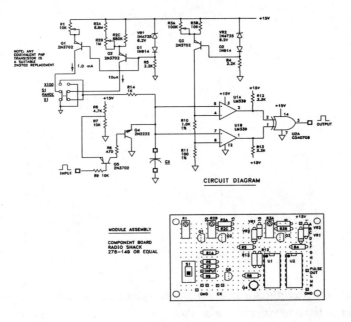

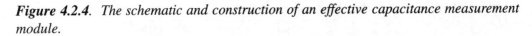

Figure 4.2.4. *The schematic and construction of an effective capacitance measurement module.*

Figure 4.2.3 illustrates the concepts. The collector of a PNP transistor appears as a high source resistance. This makes it an excellent charging source. Rather than switching the transistor on and off it is convenient to shunt the capacitor with a fast action shorting switch to ground. The switch is opened to initiate the measurement.

Two comparators monitor the capacitor voltage. Two are required because the shunt switch has a small standoff voltage. The first comparator is switched on at this value, which is set by a resistor from the minus input to ground. The second comparator senses the 1.0 volt across the capacitor by comparing it with a reference at its minus input in series with that of the first comparator. The two resistors are driven by another constant current source of precisely one milliampere.

With this arrangement:

$$C = I \bullet (t1\text{-}t2)/(Vc1\text{-}Vc2)$$

— which reduces to:

$$C = I \bullet (t1\text{-}t2)$$

— with the voltage difference correctly adjusted.

Figure 4.2.4 shows the complete circuit. An exclusive-OR gate monitors the two comparator outputs. The charging time can be read from an oscilloscope trace or a period counter. One microsecond corresponds to one picofarad of capacitance.

A single pulse can be obtained from the module of section 4.2.7. If a scope trace is used a square wave frequency source provides better readability.

4.2.4. A 1-Second Clock

This is about the simplest one-second oscillator circuit I know of in requiring but three ICs: a CMOS ICL7555 in the astable mode, a CD4040 12-stage binary divider, and one-half a CD4082.

So how does it work?

Consider the following division table:

Stage n	2n	Stage n	2n
12	4096	6	64
11	2048	5	32
10	1024	4	16
9	512	3	8
8	256	2	4
7	128	1	2

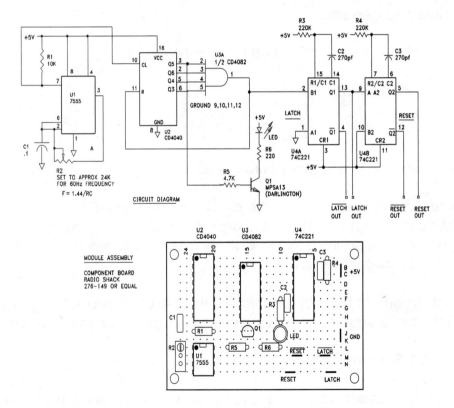

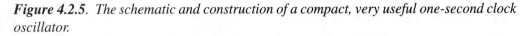

Figure 4.2.5. *The schematic and construction of a compact, very useful one-second clock oscillator.*

I have a debt to *Motorola* in that their CMOS databook shows the CD4040 can be used as a one minute oscillator by summing the Q12, Q11, Q10, and Q5 outputs in a 4-input AND logic gate. So by extension — division by 60 — summing the outputs Q6, Q5, Q4, and Q3: 3600/60 = 60. I know the sum is 7200, but the instant, so to speak, that Q12 goes high the gate responds. Another complete cycle of 60 inputs is required to drop the gate back down. Of course, this doesn't happen because the counter is immediately reset and the sequence begins all over again.

The circuit schematic and a construction model are shown in *Figure 4.2.5*. The dual monostable provides latch and reset pulses. Bringing all four mono outputs to terminal posts will accommodate about any counting circuit we might be using. The LED is nice if for no other reason than to let us know the circuit is doing its job.

If you have constructed or just browsed through the *100 Minute Timer/Stop Watch* schematic (*Figure 3.7.1*) you have seen this circuit at work in both clock modes.

4.2.5. A 4-Digit LED Display

How many times have we longed for a simple, easy to build, reliable, low frequency counter? Over the past several years I have thrown together a variety of circuits ranging from four to six digits. All used individual BCD counters, decoder/drivers, clocks, transistor switches, and the display digits.

Figure 4.2.6 is the schematic and the assembly of the end result — a basic four-digit counter. Whoa now! I hear you exclaim — there's no clock. Well, you just haven't read the preceding section, which is just the right clock for this.

The heart of the assembly is the 74C925 4-digit counter with multiplexed 7-segment output driver. The display is a compatible 4-digit common cathode assembly, *Panasonic* P/N LR513RK (*Digi-Key* P343-ND). These two items with NPN switches and resistors and a mounting board comprise the whole thing.

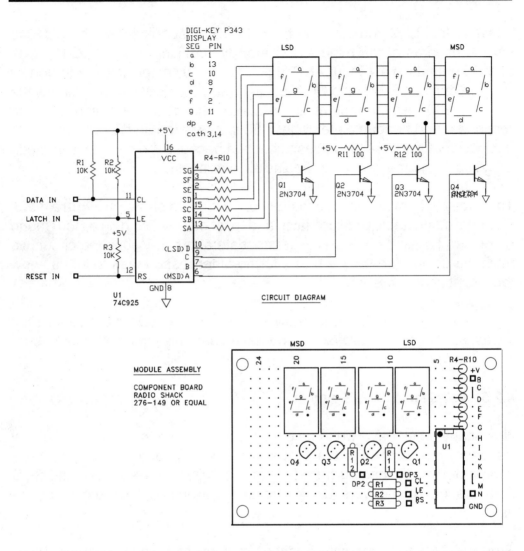

Figure 4.2.6. *The schematic and construction details of a compact, very handy LED display module.*

A description of the 74C925 is found in Chapter 5, section 5.3.7.

4.2.6. A 3½-Digit LCD Panel Meter

My original plan for this section was a four-digit display along the lines of the LED of the preceding section. Well, let me tell you: first of all, obtaining information on a display took some doing. Then when I obtained a couple to work with I discovered there is even more doing. Building up a breadboard with one of these is fraught with hidden hazards. Everything in the layout seems to be critical. And since these were not multiplexed I had jumpers everywhere.

So I got smart and bought a 3½-digit panel meter. The cost came out to less than the sum I had spent on the display and the ICL7106 converter — the converter, by the way, that is used with the panel meter.

Some specifications on the PM-128[1] panel meter are:

- Maximum input: 199.9 mV DC. With appropriate divider can measure up to 500 volts.
- Maximum display: 1999 counts (3½ digits) with auto polarity indication.
- Measuring method: Dual-slope integration A/D conversion.
- Reading rate time: 2 to 3 readings per second.
- Accuracy: ±0.5% at normal room temperature. (Will be affected by whatever divider network we might use.)
- Power: 1 mA DC.
- Decimal point selectable with wire jumper.
- Supply voltage: 9 to 12 VDC.
- Dimensions: 68 mm by 44 mm.

My *Harris Data Acquisition Databook* offered an evaluation kit for their converters, but when I tried to buy one was told they no longer offer it. Well, the circuit board on the panel meter sure resembles it. Very easy to use, just connect the resistors and jumpers as shown in Figure 4.2.6-1. It took a couple of hours to work up the assembly. Pop a nine volt battery in the holder, and we're off and running with a very convenient benchtop 20 volt DVM. A word of caution: my unit was supplied with plastic nuts for mounting. They will *melt.* And they are metric, so keep soldering irons at a distance.

4.2.7. A Bounce-Free Switch with Dual Mono Outputs

If you construct the capacitance measurement module from section 4.2.3, you will need a source for a single positive-going pulse — this is just the

1. JDR Microdevices.

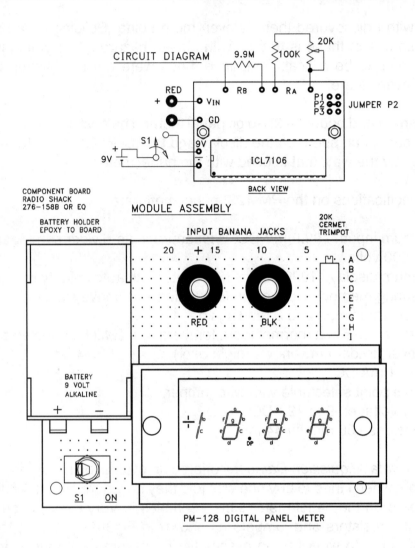

Figure 4.2.7. Circuitry for a ± 20V DC DVM based on 3-1/2 digit LCD panel meter.

module for it. Of course, you might want to use it for other projects too, right?

Figure 4.2.8 shows circuit and construction details. Switch contacts have a way of bouncing when changing state. The resulting train of pulses can have an undesirable effect, depending on the circuit. So this module em- ploys a NAND latch to obtain a good clean transition. The output can be

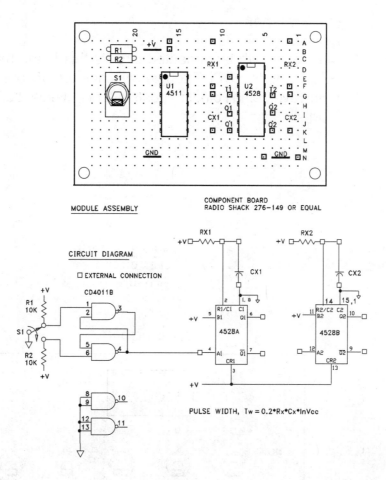

Figure 4.2.8. *The schematic and construction of a bounceless switch and a dual mono-stable, all with independent outputs. The two mono halves may be driven independently or in series for maximum versatility. The timing equation for the CD4528 with Cx > 0.01 mF is Tw = 0.2R x C x Ln (Vdd-Vss).*

used externally by provision of a binding post. It also triggers a CD4528 dual monostable. Small binding posts provide for timing resistor and capacitor connections. The timing equation is given with the figure caption.

I have found this to be a useful accessory to have at hand, and I suspect you will also.

4.2.8. Ten Toggle Switches

If you are experimenting with bus-oriented circuits this is the module for you. The eight SPDT and two DPDT toggle switches are uncommitted for maximum utility. However, small binding posts offer convenient connection points to your circuits.

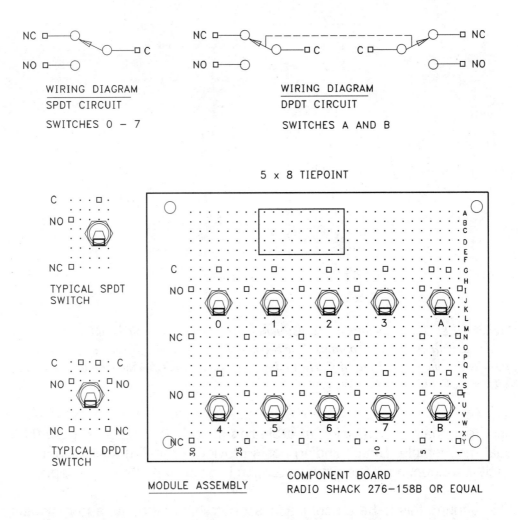

Figure 4.2.9. The schematic and construction of a toggle switch cluster module.

The item identified as *5 x 8 tiepoint* is a small matrix block that accepts #22 wire. I put it on as an accessory for connecting things like pull-up resistors and such. (*Figure 4.2.9.*)

4.2.9. Variable Clock Oscillator with Four LEDs

Figure 4.2.10 shows the circuitry and construction of a TTL or CMOS 555-based derived clock oscillator and four independent LED displays. From the schematics we see the device used may be the TTL 555 or the CMOS 7555 timer. The CMOS 7555 is a direct replacement for the TTL 555 in most applications.

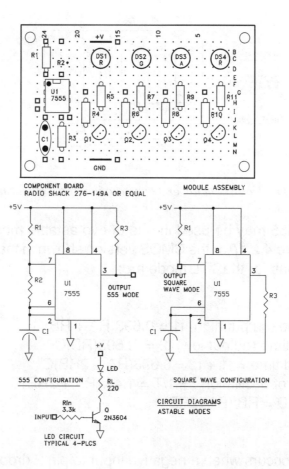

Figure 4.2.10. The schematic and construction of a clock oscillator and four independent LED displays.

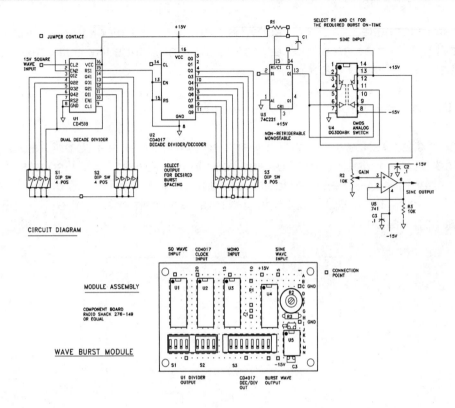

Figure 4.2.11. *The schematic and construction of a waveform burst module.*

The CMOS 7555 may be used in one of two astable modes: the TTL 555 shown in **Figure 4.2.10** or the CMOS version shown in the (b) portion of the figure. Equations for the TTL mode are:

Astable:
  Charge time (output high) $t1 = 0.693(RA + RB)C$
  Discharge time (output low) $t2 = 0.693(RB)C$
  Total period time $= t1 + t2 = 0.693(RA + 2RB)C$
  Frequency of oscillation $f = 1/T = 1.44/(RA + 2RB)C$
  Duty cycle $D = RB/(RA+2RB)$

Monostable:
  Triggering occurs when a negative input to pin 2 drops to 1/3 VCC.
  The output high time is $t = 1.1 \cdot RA \cdot C$.
  The device is not retriggerable.

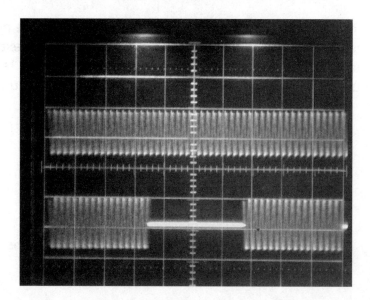

Photo 4.2.2. *A sine wave example using the burst module.* *Upper trace:* *generator frequency 5220 Hz.* *Lower trace:* *burst pattern.* *Vertical scale:* *10V/div, both traces.* *Horizontal scale:* *1 mS/div.* *External trigger from dip switch 2, position 1.*

For the CMOS mode, the monostable timing is the same as for the TTL MODE. For the astable mode, the output is a square wave with a frequency f = 1.44/R • C.

It is apparent which is the easiest to use when it comes to working with the equations.

This is a useful module for setting up a quickly needed clock source. Connection points to external circuits make a versatile bench unit. Red, green and yellow LEDs frequently aid interpretation of circuit behavior. The LEDs are helpful in tracking the progression of data through a slowly moving sequence.

4.2.10. A Waveform Burst Module

Figure 4.2.11 shows the circuitry and construction of a versatile *waveform burst* module. The waveform is typically a sine wave, but doesn't have to be. The burst timing and number of cycles are determined by the divider and divider/decoder ICs and the dip switches. A signal source and oscilloscope are necessary companions. This is a good module for exploring behavior of filters, active and passive, also for investigating the response of audio components such as speaker and amplifier responses to sudden bursts of signal.

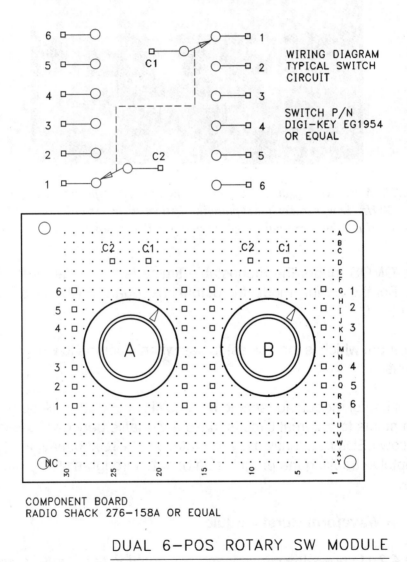

DUAL 6-POS ROTARY SW MODULE

Figure 4.2.12. The schematic and construction of a dual 6-position, 2-pole rotary switch module.

A burst pattern is shown in **Photo 4.2.2.** In this sequence the burst frequency is 5200 KHz. The control source is the square wave output from the square-triangle-sine generator of Chapter 3, section 4.

The burst module is an excellent companion to the *square-triangle-sine wave generator* described in Chapter 3.4. Applications include exploring amplifier, speaker, and passive/active filter responses.

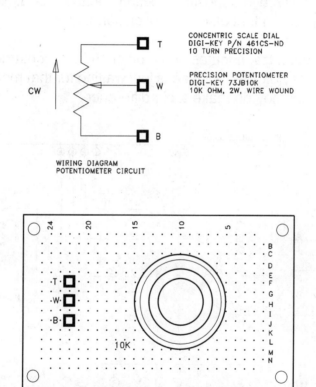

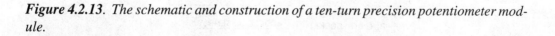

Figure 4.2.13. The schematic and construction of a ten-turn precision potentiometer module.

4.2.11. A Couple of Rotary Switches

From time to time, I have found the need to include a rotary switch in a breadboard. A loose switch with a tangle of clip leads turned out to be more of a hazard than a help. Hence this module. It's not used all that often, but it's handy when needed.

Figure 4.2.12 shows how I did the assembly and wiring. Note the two sections of the switch wire in opposite directions on the board. Not shown is a four-pole, five-position that also finds occasional use.

I found it takes but a few minutes and a bit of effort to construct modules of this kind. True, they often sit for weeks just waiting for that moment of glory, but when it comes they do make a big difference.

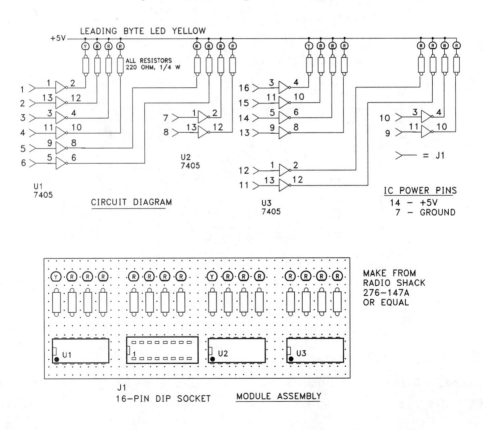

Figure 4.2.14. This module features sixteen mini LEDs clustered in groups of four.

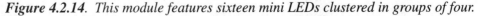

4.2.12. A Precision Potentiometer

It's pretty hard to use a ten-turn precision dial with a floating ten-turn poten-tiometer. It really needs a surface to ride on. *Figure 4.2.13* shows the schematic and construction of a *ten-turn precision potentiometer* module.

This module is very useful for analog value settings. The precision dial is particularly helpful when you don't have a digital voltmeter or an oscillo-scope for measurements. It simplifies data taking and returning to previous values. I have found a 10K pot fills the majority of my needs.

4.2.13. An IC Inverter Driven LED Bus Display

In 1976, I bought a KIM-1 single board computer. The I/O ports were great, but keeping track of the highs and lows was a challenge till I constructed the LED assembly you see in *Figure 4.2.14*.

This module features *sixteen mini LEDs* clustered in groups of four for ease of readability. Use of a yellow LED for the high bit is surprisingly helpful for keeping track of which end is the most significant. The 74HC05/74HCT05 will work equally well here. The 16-bit socket is for ribbon cable DIP assemblies. Sixteen individual clip leads can get pretty messy.

4.2.14. A Low-Voltage Detection Module

If you've read through chapter three, or better yet, done some of the construction, this module looks familiar. My purpose in including it here is to present a complete assembly for stand-alone uses. Although the original circuit was to monitor a battery powered instrument it has value wherever a drop in voltage needs to be detected. The addition of a buzzer is easily accomplished.

The component values shown in *Figure 4.2.15* are for a six volt battery with the low voltage to be set for about 4.3 volts. The low voltage detection is set using an external variable voltage source. Potentiometer R2A is adjusted to light the amber LED as the power supply is reduced from the six volt level.

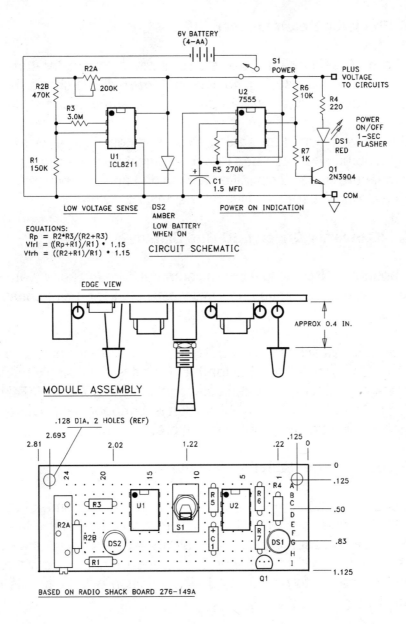

Figure 4.2.15. This module was constructed to monitor a battery powered instrument where it is desirable to detect the voltage at which the battery should be replaced.

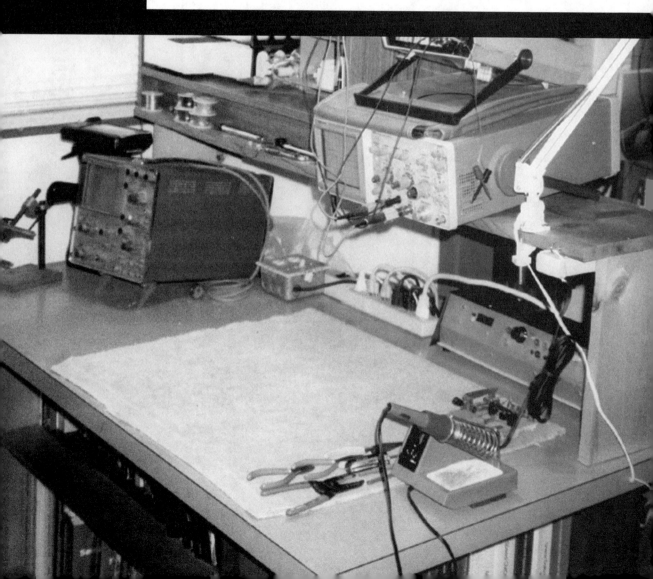

CHAPTER 5

CIRCUIT DATA BASICS FOR US TO BUILD UPON

CHAPTER 5
CIRCUIT DATA BASICS FOR US TO BUILD UPON

5.1. DISCRETES

Most of what we have been reading to this point is giving us ideas and suggestions on what we might do for ourselves — thoughts on setting up our own little facility, on instrumentation and breadboarding and modules. Now comes the topping off with insights on a portion of the fabulous variety of discretes, analog and digital integrated circuits, and the vast array of optoelectronic devices just waiting for us to explore.

I wish I could shoehorn ever so many more devices into this chapter, but reality limits the choices. In part, the descriptions relate to circuits I have worked with; and in part, to those I believe to be of value to you as hobbyist, experimenter or student.

When we dine out, we are provided a menu from which to make our choices. We are not going to order every item (at least not in a single evening), but we will peruse the list and toy with the possibilities. If we really enjoy what the restaurant has to offer, we will return from time to time for other favorites.

This is the approach best taken with this chapter.

Of course, space also imposes limitations on how much description can be given. Use the manufacturers' data books to follow up with those details important to your project.

5.1.1. Rectifier and Voltage Reference Diodes

Basic Principles

The movement of electrical charges within a semiconductor material takes place in solid metal — a material processed to standards of crystal structure and purity so rigid in their demands for perfection as to be almost beyond comprehension.

When a pure, mechanically perfect crystal of silicon is kept at zero degrees Kelvin (0°K, -273°C), it acts as an ideal insulator. The material is unable to conduct electricity because the atoms in the crystal lattice structure have formed electron sharing bonds with their neighbors. The bonds are stable and there are no free electrons for the transport of charge through the material.

As the temperature is raised the heat energy that becomes available causes an occasional bond to break. This releases an electron which can become a charge carrier. In addition, the atom from the broken bond is also free to migrate. This broken bond carries one positive charge. It is called a *hole.*

So as the temperature rises the intrinsic (pure) material changes its nature from an insulator to a semiconductor. The density of the free carriers is dependent on the semiconductor material and the temperature. For the mathematically inclined the relationship is given by:

$$ni^2 = pi^2 = nipi = AoT^3 \exp(-Egq/kT)$$

— where *ni, pi* are the respective electron and hole densities; q is the charge on the electron, 1.6×10^{-19} coulomb; *k* is Boltzman's Constant, 1.38×10^{-23} joule/°K, *T* is the absolute temperature in °K; *Ao* is a constant; and *Eg* is the energy-gap voltage, a material constant. For silicon at 300°K, (°K=°C + 273), *Eg* = 1.106 volts; *ni = pi* = $2 \cdot 10^{10}/cm^3$.

The two mechanisms by which the charge carriers carry a current are (1) drift of the carriers under the influence of an electric field, and (2) diffusion under the influence of a density gradient. These are called *drift current* and *diffusion current*, respectively.

The drift current is an *Ohm's law* effect for resistive materials. This is what occurs when we connect a battery in our circuit. In the absence of an applied electric field the diffusion current takes place as the charge carriers move about randomly, experiencing collisions with the crystal lattice atoms. Since the number of positive (hole) and negative (electron) charges is equal there is no net electric charge.

Intrinsic material in itself is not a useful substance for working semiconductor devices unless we are in need of temperature sensitive resistors. What is needed is a means of changing the relative densities of the negative and positive charges so as to create an inequality. This is accomplished by a procedure known as *doping.*

If the number of electrons is increased over the holes the material is said to be *N-type.* When the hole density predominates we have *P-type* material. Various manufacturing processes have been developed for the doping. The objective is to create a PN junction within adjoining regions of the material. *Figure 5.1.1* (a) is a basic PN junction — a diode, in workaday terms. The actual number of impurity atoms needed for the doping is minute.

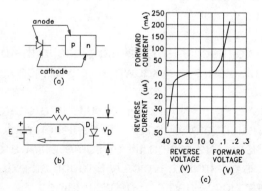

Figure 5.1.1. Junction diode characteristics

In the absence of an external field the charges will migrate throughout the material. The introduction of an external field will either draw the carriers away from the junction, in which case the device is *reverse biased,* or toward the junction, in which the device is *forward biased.*

Figure 5.1.1 (b) illustrates the direction of current flow across the junction with an external positive voltage. The (c) portion of the figure shows the conduction characteristics with positive and negative potentials across the diode. (Actual diode values will vary, the numbers shown are for illustration only.)

This basic junction, while having a variety of material and doping variations it is true, is the basis for all the semiconductors so widely in use today.

The Zener Diode

We saw in *Figure 5.1.1* (b) that a reverse voltage across the PN junction is resisted up to some point at which a small conduction begins. The reverse current will slowly increase up to some value, not shown in the diagram, where the junction cannot sustain the potential and a total collapse takes place.

The curvature at which the reverse flow begins its increase is the *current knee.* It is desirable that this be as abrupt as possible. It is also desirable that the current remain relatively constant over the allowable range of the current reversal. Under these conditions the device functions very nicely as a constant voltage regulator.

There are actually two mechanisms taking place in the breakdown region to maintain the effect: zener breakdown and avalanche.

The creation of a reversed PN junction results in a charge-free region on either side of the junction referred to as the depletion region. Increasing the reverse voltage widens this region till at some point a breakdown takes place. If the current is controlled, typically through an external series resistance, this is nondestructive and can be maintained.

An avalanche is induced by collisions between charge carriers that induce further collisions by the energy of the collisions. A large current may result with little change in voltage.

The properties of interest are the breakdown voltage, its *ON impedance,* the *electrical noise* produced, and its *temperature characteristics.* Over the years a variety of devices have evolved that will satisfy almost any requirement.

As an experimenter, you may enjoy knowing the base-emitter diode of a silicon transistor can be used in the breakdown mode as a voltage reference. The actual voltage is dependent on the manufacturing process and varies over transistor types. Typical breakdowns range from four or five volts to fifteen or more for small signal devices.

The Precision Voltage Reference

A class of diodes have been developed as precision voltage references with extremely low temperature coefficients (TC). With these advantage is taken of the differing thermal performance of forward and reverse biased diodes. A forward biased junction has a negative coefficient of approximately 2 mV/°C. Below five volts reversed junctions also experience a negative coefficient, which slowly becomes positive at higher voltages. Through careful selection a combination of forward and reverse biased junctions yield a device with a coefficient very close to zero at a specified current. The Motorola 1N821 - 1N829A featured TCs ranging from a high of .01%/ °C to a low .0005%/°C. The zener voltage for this series is 6.2 volts, $\pm$ 5% at a current of 7.5 mA. The 1N935 - 1N939B provided a constant 9.0 volts with the same current and temperature specifications.

As a hobbyist, we can construct our own quasi-low coefficient diodes. Find a PNP transistor with a base-emitter breakdown close to six volts and operate it with the collector-base diode in series. The forward biased diode contributes a -2mV/°C coefficient, partially canceling that of the base-emitter. I did this years ago using the Fairchild 2N3638. The results were surprising good with a diode current of about ten microamperes.

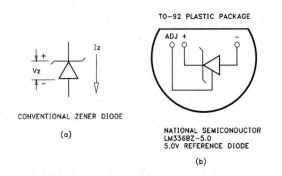

Figure 5.1.2. Zener diode and precision voltage reference symbols.

Several manufacturers have developed integrated circuit references with high levels of performance. An example is the National Semiconductor series of precision references. Examples are the LM113/313, 1.22 volts; the LM185-1.2/LM385-1.2, 1.235 volts; the LM136-2.5/LM336-2.5, 2.49 volts, and up to the LM199 - LM3999, at 6.95 volts. These represent considerable improvement over the previous compensated zeners in allowing a greater range of operating current. **Figure 5.1.2** illustrates the symbols for a conventional zener diode and a precision voltage reference. **Table 5.1.1.** contains some diode definitions.

5.1.2. The Bipolar Transistor

When the transistor first came into use back in the late fifties they were constructed of germanium with specifications that would be totally unacceptable in today's engineering; but at the time their performance was viewed as near miraculous. (To engineers and technicians accustomed to vacuum tubes, other adjectives were applied from time to time.) By the end of the sixties, silicon and the integrated circuit (IC) had taken over and the solid-state boom was in full swing.

This would seem, of course, to lead to a greatly diminished usage of the transistor, but it is by no means a has-been device. Just a glance through any manufacturer's small-signal and bipolar power transistor data books offers convincing proof that the transistor is alive and prospering.

Diode Terminology

Symbol	Definition	Units
VRRM	Reverse Breakdown Voltage	Volts
IR	Reverse Breakdown Current	nA
VF	Forward Voltage Drop	Volts
C	Diode Capacitance	pF
trr	Reverse Recovery Time	nS

Zener Diode Terminology

Symbol	Definition	Units
VZ	Nominal Zener Value	Volts
IZ	@ this Current	mA
ZZ	Dynamic Diode Resistance	Ohms
TC	Zener Temperature Coefficient	%/°C
TA	The Reference Ambient Temp.	25°C

Glass Packaged Diode Classifications

Type	Examples
Computer	1N914, 1N916, 1N4148
Low Leakage	1N456 - 1N459, 1N482B
High Voltage	1N625, 1N629, 1N3070
Gen. Purpose	1N461, 1N660
Zener, 500mW	1N746A - 1N973A, 3.3 - 33 volts
Zener, 500mW	1N5226B - 1N5757B, 3.3 - 33 volts
Zener, 1000mW	1N4728A - 1N4752A, 3.3 - 33 volts

Table 5.1.1. Some diode definitions.

Transistor Terminology and Characteristics

In the beginning, there were the NPN and the PNP structures, as of course there still are. *Figure 5.1.3* illustrates the basics of the NPN transistor. The diode concept shown is not a literal reality — you cannot make your own transistors in this fashion — but it does illustrate the principles of transistor action. The PNP has the same structure: just invert the diodes and voltages. The use of negative power supplies, primarily for the early era PNP requirements, has been greatly diminished by the IC's positive voltage requirement. So this discussion considers positive power only. To visualize negative power for the PNP, flip it over to show its emitter at the positive level.

The transistor is a current-operated device. With reference to **Figure 5.1.3**, we observe that the base-emitter diode is forward biased while the base-collector is reverse biased. Without going into details on why, this is the mechanism that provides the amplification.

We see from the circuit the emitter current, IE, to be the sum of the base, IB, and collector, IC, currents. The current gain is basically the collector/base ratio. This ratio is called the DC ß (Beta). A high Beta is a figure of merit.

Figure 5.1.4 shows the device symbology and a variety of package concepts that have evolved. The *Darlington* connection shown is one we can make for ourselves. But they have come down so low in price it hardly pays to do so.

Figure 5.1.5 illustrates modes of transistor operation. In essence there are two: linear and nonlinear. We find linear applications in the analog world; nonlinear as switches and logic devices.

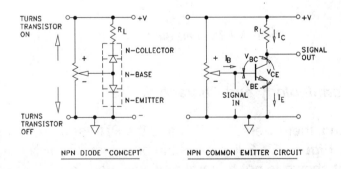

Figure 5.1.3. *Bipolar transistor basic concepts. Though the NPN only is shown these apply to the PNP model also with diodes and voltages inverted.*

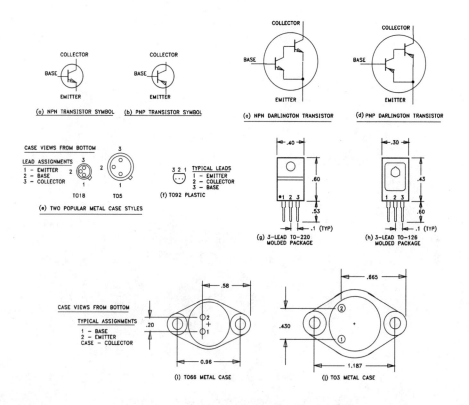

Figure 5.1.4. *Bipolar transistor symbology and an and assortment of case styles that have evolved.*

We see that in the *common emitter* circuit, the base is but one diode potential above common. So in general, we need a resistance between the base input and the signal source. An advantage of the emitter follower is current gain along with amplification of the emitter circuit resistance as seen when looking in at the base. We trade the loss of voltage gain for reduced loading on the source. Beyond that, the source impedance looking back into the emitter is also greatly reduced. There is a third configuration, *common base,* not shown in the drawing. As the name implies, the base becomes the ground reference, with the signal input to the emitter. It features a very low input impedance and very high output impedance.

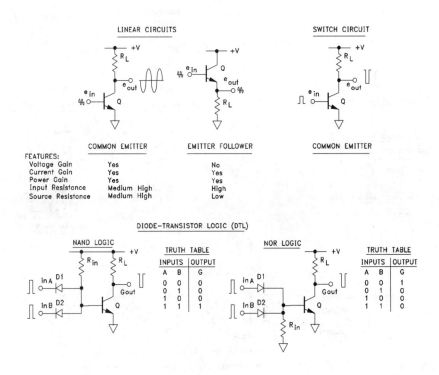

Figure 5.1.5. *Some transistor applications.*

The plot of **Figure 5.1.6** describes typical common-emitter collector characteristics. The flat appearance in the active region arises from the high collector source resistance. The nearly vertical line from which the characteristic lines depart defines the *collector saturation* region. In this region the collector to emitter parameter appears to be a low value resistance shunted by capacitance; the intervening voltage, VCE, is very low. This region is unsuitable for linear use, but highly desirable for switching applications.

The slanting line is referred to as the *collector load line.* The dots indicate operating points. Suppose we consider the center dot — located at 2 mA and 8 VDC as the quiescent operating point — to be collector values with no AC signal input. The other two dots represent the maximum and minimum signal swing peaks. If we allow the collector voltage to go beyond these points there will be excessive distortion.

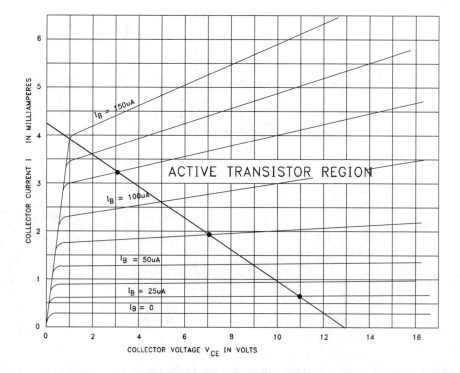

Figure 5.1.6. *Collector characteristics for a typical bipolar unit.*

We mostly use discrete transistors as switches, the IC being so much better qualified for more demanding tasks. The primary requirement as a switch is a fairly high base current drive. When used as a switch we want to drive the collector voltage to as close to ground as possible, and as swiftly as possible, to minimize heat losses in the collector. But the high base current demanded lowers the transistor Beta; it is best to assume a drop to about ten. We want to be sure whatever is sourcing the switch can provide the required drive. This is where the *Darlington* connection can really shine.

There is an upper limit on speed determined by junction capacitances and for electron and hole diffusion times. These are defined by manufacturing processes. Where the speed is critical the manufacturer's data must be consulted.

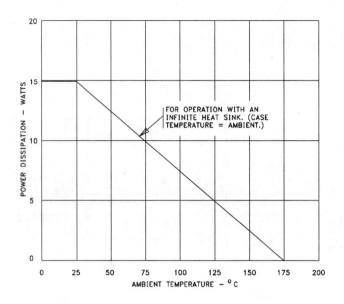

Figure 5.1.7. Power transistor derating curve.

Heat is the primary cause of semiconductor failures in service. Heat buildup becomes very significant in power applications. Only a glimpse of the requirements can be included here. The manufacturer's data must be consulted for the selection of a suitable transistor, and the operations environment given a thorough review to assure long-term reliability.

Figure 5.1.7 is a typical heat transfer curve. It simply defines the allowable device dissipation for a given ambient, i.e, case, temperature.

The flow of heat in a thermal circuit is very analogous to current in an electrical circuit. The heat source, say the transistor collector junction, is the battery. Heat flows from points of high temperature to the thermal common, ultimately the surrounding atmosphere. Thermal resistances arise at boundaries, with resultant temperature drops. The transfer of heat from a small metal surface to the atmosphere represents a very large thermal resistance. We reduce this by mounting the transistor on a *heat sink*, a metallic structure having a large surface. This why heat sinks have fins - they in-

TJ	Junction Temperature
TA	Ambient Temperature
RΘJC	Junction-to-Case thermal resistance.
RΘCA	Case-to-Ambient thermal resistance.
RΘCS	Case-to-Heat Sink thermal resistance.
RΘSA	Heat Sink-to-Ambient thermal resistance.
RΘJA	Junction-to-Ambient thermal resistance, heat sink included.
	= RΘJC + RΘCA + RΘCS + RΘJA

Table 5.1.2. Thermal relationshups between the semiconductor junction and ambient (room) temperature.

crease the radiating surface. Black is the preferred color for heat transfer. We generally think of black as absorbing heat, and this is true; but the best absorbers are also the best radiators. *Table 5.1.2* defines the heat flow path with its thermal interfaces.

5.1.3. The Field Effect Transistor (JFET, MOSFET, CMOS)

This section provides insight into the construction and usage of a unique and valuable circuit element. Basic characteristics, terminology, establishing an operating point, and some illustrative usage examples are described.

The Junction Field Effect Transistor (JFET)

The JFET is sometimes called a "unipolar" transistor. This because, in contrast to its bipolar cousin, the current transport is by majority carriers. Its makeup is relatively uncomplicated, consisting of a piece of high-resistivity silicon to provide a "channel" for the carrier conduction. A reverse biased PN junction is formed along the channel edge. This yields a voltage gate for control of the current flow through the channel. In a sense we can think of the gate signal as modulating the channel carriers. The back-biased gate offers a high resistance to the signal source. This is a sharp contrast to the low input offered at the base of a bipolar transistor. The FET is a voltage controlled device in contrast to the current control of the bipolar transistor.

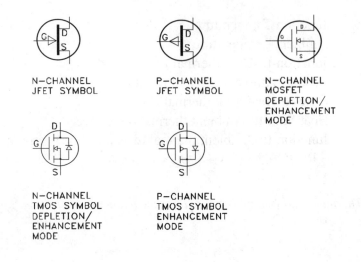

N-CHANNEL
JFET SYMBOL

P-CHANNEL
JFET SYMBOL

N-CHANNEL
MOSFET
DEPLETION/
ENHANCEMENT
MODE

N-CHANNEL
TMOS SYMBOL
DEPLETION/
ENHANCEMENT
MODE

P-CHANNEL
TMOS SYMBOL
ENHANCEMENT
MODE

Figure 5.1.8. FET electrical symbols.

If the channel is doped with a donor (electron) impurity, N-type material results. Similarly, if doped with an acceptor (hole) impurity, the channel is P-type. The N- and P-channel JFET symbols are identified in *Figure 5.1.8*. N-channel devices possess a higher conductivity than P-channel, thus tending to be more efficient. An ohmic contact enables external termination connections. The three terminals are named *gate* (G), *drain* (D), and *source* (S). The drain corresponds to the bipolar collector, the source to the emitter. Typically the drain and source are symmetrical and may be interchanged.

Device packaging is similar to that for bipolar transistors, as shown in *Figure 5.1.4*.

The Metal-Oxide Semiconductor Field Effect Transistor (MOSFET)

These devices differ from the junction type by use of an insulating oxide layer isolating the channel from the gate. The MOSFET functioning is based on the fact that there is no real requirement for an externally connected PN junction in physical contact with the channel to enable channel current control. In these devices a metallic or polysilicon gate is isolated from the channel by a thin layer of silicon dioxide, otherwise known as glass. Physical processes form the drain-source conduction path. MOSFET drawing symbols are included in *Figure 5.1.8*.

MOS devices feature very high input resistances and are therefore sensitive to static electricity. Although many devices feature internal protection they should always be treated with safe handling precautions.

MOSFETs are constructed to operate in one of two channel conduction modes: *depletion* or *enhancement*. A depletion mode device offers low channel resistance in the absence of gate bias; the bias limits channel current by forcing a reduction in its cross section thereby decreasing its conductance. The channel of an enhancement device is nonconducting in the absence of a gate signal; the gate bias enables channel current flow. JFETs are inherently depletion mode; MOSFETs may be of either mode; some types may be designated as both, though in general depletion/enhancement types are operated in the depletion mode.

MOSFETs have been given a variety of names by various manufacturers, among them HEXFET[1], TMOS, and DMOS. These arise from differences in their construction to favor certain attributes.

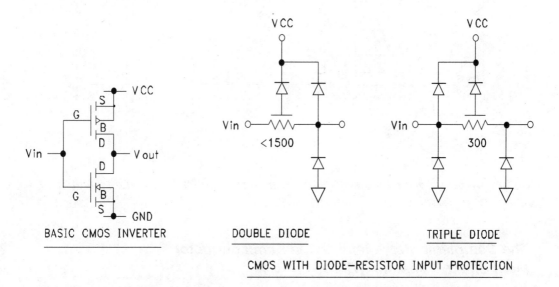

BASIC CMOS INVERTER   DOUBLE DIODE   TRIPLE DIODE

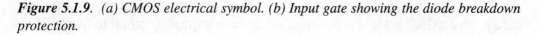

CMOS WITH DIODE–RESISTOR INPUT PROTECTION

Figure 5.1.9. (a) CMOS electrical symbol. (b) Input gate showing the diode breakdown protection.

1. *HEXFET is the trademark for International Rectifier power MOSFETs. DMOS is a product of Siliconix Incorporated. TMOS is a product of Motorola Incorporated.*

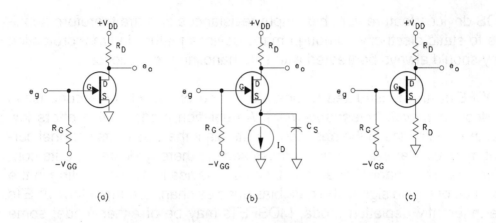

Figure 5.1.10. *FET biasing examples. (a) Fixed voltage. (b) Constant current. (c) Self bias.*

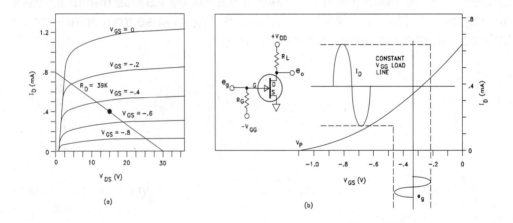

Figure 5.1.11. *Example of a constant-voltage bias amplifier based on a typical JFET's output characteristics, (a), and transfer curve, (b).*

The Complementary Metal-Oxide Semiconductor Field Effect Transistor (CMOS)

An ideal logic family should dissipate no power, have a zero propagation delay, controlled rise and fall times, and noise immunity equal to 50% of the logic swing. CMOS properties are such that these devices approach the ideals.

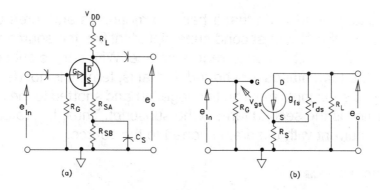

Figure 5.1.12. (a) Common source feedback with source resistance feedback. (b) The simplified equivalent circuit.

The basic CMOS construction is an inverter, as shown in **Figure 5.1.9**(a). The circuit consists of two enhancement mode transistors. The upper is a P-channel type, the lower N-channel. Some manufacturers designate the DC power as VDD and VSS for positive and ground, others the more familiar VCC and GROUND.

Actually there are no discrete devices your author knows of; it is included here as an MOS related technology. Most logic ICs now include static protection as shown in **Figure 5.1.9**(b).

FET Terminology and Parameters

Major parameters include:

- I_{DSS}: Drain current with the gate shorted to the source.
- $V_{GS(OFF)}$: Gate-source cutoff voltage.
- I_{GSS}: Gate-to-source current with the drain shorted to the source.
- $V_{(BR)GSS}$: Gate-to-source breakdown voltage with the drain shorted to the source.
- g_{fs}: Common-source forward transconductance.
- C_{gs}: Gate-source capacitance.
- C_{gd}: Gate-drain capacitance.
- V_p: Any combination of VGS and VDS forcing channel resistance, rDS, to infinity. Referred to as the "constant current" region.

The "S" subscript is tricky in that it has two meanings and three possible uses. An "S" for the first or second subscript identifies the source terminal as a node point for voltage or current reference. With a triple subscript, an "S" for the third term stands for shorted — that is, terminals not designated by the first two subscripts must be tied together and shorted to the common terminal, which is the second term in the subscript. Thus, I_{GSS} refers to the gate-source current with the drain shorted to the source.

FET Characteristics

FETs offer certain advantages over bipolar devices. These arise from their construction and modes of operation. Included are:

- Low noise.
- "Zero offset" on resistance.
- No thermal runaway.
- Low distortion and intermodulation distortion.
- High input impedance at low frequencies.
- Very high dynamic range.
- Junction capacitance independent of device current.

The input resistance is simply that of a reverse biased PN junction for the JFET, a very high leakage resistance with the MOSFET. This may exceed a trillion ohms. It enjoys a dynamic range in excess of 100 dB. The low noise and distortion enables amplification of a broad range of signal levels. There exists a JFET bias point at which a zero temperature coefficient is exhibited. (However, I have observed significant drifts when improperly biased.)

JFET Biasing

Correct biasing is essential for reducing the device sensitivity to undesired characteristics. In some respects it is more critical than for bipolar transistors. I cannot begin to describe all the available options. What follows will provide some insight. For further needs it will be necessary to refer to a data manual, such as Siliconix, which is very descriptive.

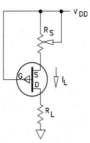

Figure 5.1.13. *Constant current source using a P-channel FET. This circuit gives excellent results in the microampere current range.*

The preferred biasing is frequently a combination of constant voltage and self-biasing. This discussion is limited to the common source mode as it is that most commonly used. We can think of it as equivalent to the bipolar common emitter mode.

There are three basic biasing circuits suitable for fixing the FET's operating-point (Q-point).

- Constant-voltage bias, most useful for rf and video amplifiers employing small DC drain resistors.
- Constant-current bias, best suited to low drift DC amplifier uses such as source followers and source-coupled differential amplifier pairs.
- Self-bias, also called *source* or *automatic* bias — a somewhat universal approach, good for AC amplifiers.

These are illustrated in ***Figure 5.1.10***.

As a first step, we need to determine the Q-point. For this we need the output characteristics for the device we will be using. (These are given in section 7 of the *Siliconix* manual[2].) ***Figure 5.1.11*** is for a typical N-channel device. A *load line* is constructed from two points: the value of VDS on the X-axis and the $I_{D(MA)}$ value selected on the Y-axis. We can slant this line for the desired performance. The example is a Q-point V_{GSQ} = -0.4 volt at V_{DS} = 15 volts. This yields an RD of 39K.

2. *Siliconix Incorporated, Low Power Discretes Databook. Siliconix Corporation, 1980.*

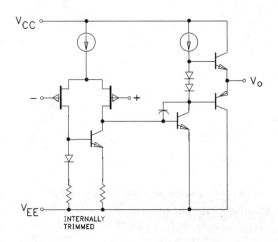

Figure 5.1.14. *Integrated circuit JFET op amp (a) simplified schematic, (b) typical user connection.*

Figure 5.1.12 shows a common source amplifier using self-bias with source resistance feedback. Splitting the resistance and bypassing the lower improves the high frequency response.

Applications

My own most frequent use of the small signal, low power, JFET has been as a constant DC current source. It is well suited for this due to its inherently high drain impedance. *Figure 5.1.13* shows the basic circuit using a P-channel FET.

The JFET is also an excellent input for low noise amplifier inputs. I use JFET input op amps such as the LF355 and LF356 in many of my circuits. *Figure 5.1.14* is the simplified schematic of the LF442 dual JFET input amplifier. (I find the FET source circuitry rather intriguing.)

With proper biasing, the FET channel is a useful voltage controlled resistor. This attribute finds utility in analog switching applications. *Figure 5.1.15* shows one approach to an integrated switch using the Siliconix DMOS FET.

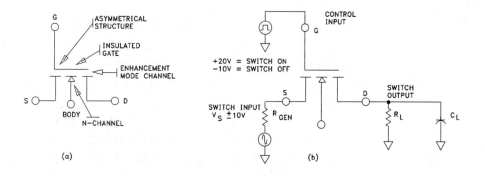

Figure 5.1.15. *DMOS FET switch. (a) Electrical symbol. (b) Configuration for a ± 10 volt analog switch.*

5.1.4. THE UNIJUNCTION TRANSISTORS (UJT, PUT)

About 1960, the General Electric Company, which was a major manufacturer of discrete semiconductors at that time, introduced a series of unijunction transistors. Recalling that the integrated circuit was still some time away in the future, they were a welcome addition to the engineer's collection of design goodies.

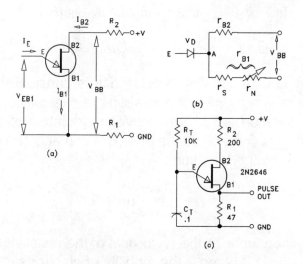

Figure 5.1.16. *(a) The unijunction transistor (UJT) electrical symbol. (b) The UJT equivalent circuit in the negative resistance region. (c) A typical oscillator application.*

IE	Emitter Current
IEO	Emitter reverse current. Measured between emitter and base-2 at a specific voltage with base-1 open.
Ip	Peak point emitter current. The maximum amount of emitter current that can flow without the UJT going into a negative resistance region
IV	Valley point emitter current. The current flowing in the emitter when the current is at the valley point.
rBB	Interbase resistance. Resistance between base-2 and base-1 measured at a specified interbase voltage.
VBB	Voltage between base-2 and base-1. Positive at base-2.
VP	Peak point emitter voltage. The maximum voltage seen at the emitter before the UJT goes into the negative resistance region.
VD	The forward voltage drop of the emitter junction.
VEB1	Emitter to base-1 voltage.
VV	Valley point emitter voltage. Occurs with a specified VB2B1.
VOB1	Base-1 peak pulse voltage, measured across a resistor in series with base-1 in a specified circuit operating as a relaxation oscillator.
n	Intrinsic standoff ratio, = VP-VD/VB2B1.
αrBB	Interbase resistance temperature coefficient.

Table 5.1.3. Unijunction transistor definitions.

The present day construction of the UJT differs considerably from those early devices, but the principles of operation are basically unchanged. The basis for their operation is the conductivity modulation of the silicon between the emitter junction and the base-1 (B1) contact. The conductivity in the region is given by the equation[1]:

$$\sigma = (\mu p^p + \mu n^n)$$

— where σ = the conductivity, the reciprocal of the resistivity, μp = the mobility of holes in silicon, and μn = the mobility of electrons in silicon.

1. Reference T. P. Sylvan, General Electric Company, Notes On the Application of the Silicon Unijunction Transistor, Applications Engineering, Syracuse, New York, 90.10 5/61.

Figure 5.1.16 (a), (b), and (c) define the important attributes of the UJT. (a) shows the electrical symbol we use on our drawings. *Table 5.1.3* defines the important UJT parameters. If we look at the equivalent circuit of (b), we see a resistance with an arrow through it, indicating that it varies with something. The something is the voltage building up on the capacitor in the oscillator circuit of (c). The diode represents the base-emitter junction. As the voltage builds up the conductivity noted in the equation increases. At some point the current flow through the emitter goes through a very abrupt increase as the conductivity takes a sudden drop. The result is a voltage pulse across resistor R1. The charge on the capacitor is quickly reduced to a very low value, the decrease lowers the emitter current, and the resistance is restored to its previous high value. The cycle then begins to repeat.

The resistance R_{B1} is the sum of the two equivalents, Rs and Rn. The total interior *base resistance* is given as $R_{BB} = R_{B1} + R_{B2}$. A voltage divider exists at point A in the equivalent circuit with a standoff ratio:

$$n = R_{B1}/(R_{B1}+R_{B2})$$

With no emitter current flowing, the voltage at point A is a small fraction of the total across R_{BB}, equivalent to $n = R_{B1}/R_{B2}$.

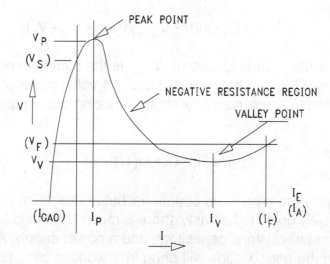

Figure 5.1.17. Static characteristics for the UJT and PUT. PUT only characteristics enclosed in parenthesis.

The real question now is how we determine the parameters of the operation cycle. The essentials are shown in the emitter characteristic curve shown in *Figure 5.1.17.* Two critical points on the curve are the *peak point* and the *valley point.* At these two points the slope of the characteristic curve is zero. The region to the left of the peak point is the *cutoff region* — the emitter diode is back-biased. The region between the peak and valley points is called the *negative resistance region.* This is where the modulation of RB1 is taking place. The region to the right of the valley point is the *saturation region.* Conduction here is by the surface and bulk hole-electron recombination.

Two external bias resistors, R2 and R1, are required. A principle variation of the peak point voltage is due to the temperature coefficient of the base-emitter diode. Proper selection of R2 can provide compensation. An approximate value for R2 is:

$$R2 \sim 0.70 \bullet R_{BB}/n \bullet V + (1-n)R1/n$$

Then, $V_P = n \bullet V$, where V is the power supply voltage.

The relaxation oscillator is the principal use made of the UJT. The period of oscillation, T, is given by:

$$T = R_T \bullet C_T \bullet \ln(V - V_{E(min)}/V - n \bullet V_{BB} - V_D)$$

— where ln is the natural logarithm. $V_{E(min)}$ is the minimum emitter voltage, approximately 0.5 $V_{E(sat)}$, between 1.2 and 2.4 volts, generally. The two resistors, *R1* and *R2,* are normally small, reducing the equation to a more reasonable:

$$T = R_T \bullet C_T \bullet \ln(1/1 - n)$$

There are many more possible equations; but at this point, unless the environment is really going to be nasty, there is much to be said for going to the bench with a few resistors, capacitors, and a power supply. A bit of experimenting with the oscilloscope will bring in a working circuit. The essential thing is to keep the two base resistors small, on the order shown in the figure. Do not load the output pulse unduly — insert a bipolar or FET driver to do the work.

The Programmable Unijunction Transistor (PUT)[2]

This device is unique in that its operation is similar to that of an SCR, but its usage is close to that of the UJT. There are many similarities between the two devices, which can be seen from a comparison of **Figures 5.1.16** and **5.1.18**. The PUT is currently the more popular device, having a greater speed and versatility. It has value as a phase controller and long duration timer.

Much of what has been said regarding the UJT applies to the PUT and is useful to an understanding of the device. The static UJT characteristic curve of **Figure 5.1.17** is equally applicable to the PUT. Where PUT terminology differs from that for the UJT is enclosed in parenthesis. **Table 5.1.4** defines the important PUT parameters.

From **Figure 5.1.18**, we see that the PUT resembles an SCR except for the anode trigger. When the gate is negative with respect to the anode the PUT will transit from a blocked to a conducting state. To enable operation as a UJT a reference voltage must be maintained at the gate. This function is performed by resistors R1 and R2 in **Figure 5.1.18** (b). The timing components, RTCT, are connected to the anode.

You might find it interesting to construct your own PUT using the PNP-NPN equivalent.

Referring back to **Figure 5.1.17**, the peak point voltage, V_p, corresponds to the reference voltage minus the gate drop. The reference is circuit controlled rather than device, which accounts for the *programmable* in its name. Although two resistors are shown in the figure any suitable voltage source will do. The *Thevenin* equivalents for the gate circuit may be expressed as:

$$V_S = R_1 \bullet V_1/(R_1+R_2)$$
$$G_S = R_1 \bullet R_2/(R_1+R_2)$$

2. *Reference Motorola Incorporated, Thyristor Device Databook, 1988.*

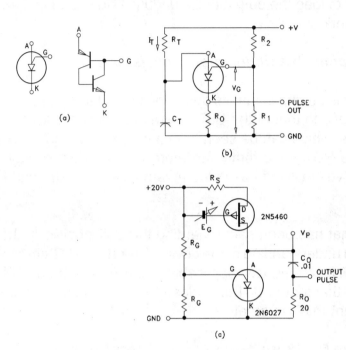

Figure 5.1.18. *(a) The programmable unijunction transistor (PUT) electrical symbol and its PNP-NPN transistor equivalent. (b) A typical oscillator circuit example. (c) A test circuit for Ip, Vp, and Iv.*

The current through the charging resistor, I_{RT}, must be greater than IP at VP to ensure switching between on and off states as an oscillator. Consult device data sheets for maximum values of IP. There is a dependency on temperature and gate characteristics, so the data should be consulted for these also.

Figure 5.1.18 (c) is a circuit for observing peak and valley characteristics. Inserting 1.5 volt batteries is the simplest way of varying gate potential, V_G.

To obtain the maximum frequency of oscillation, it is recommended to use the minimum value of capacitance, 1000 pF. Select devices and biasing for a high IV. The minimum value of timing resistance is obtained by a rule of thumb:

IL	Latching current.
Vp	Peak point voltage. VP = VT + VS, where VS is the unloaded divider voltage and VT the offset voltage.
Ip	Peak point current. With currents less than Ip the device is in the blocking state.
IV	Valley point current. A voltage minimum occurs at this point. When used as an oscillator the charging current must be less than IV at the valley point voltage, VV.
VV	Valley point voltage. The minimum voltage on the static characteristic graph.
VD	The gate diode drop.
VF	The forward voltage, that between the anode and cathode when the device is biased on.
VO	The peak output voltage. A function of VP, VF, dynamic impedance, and switching speed.
tr	The output pulse rise time. Typically about 40 nS, comparable to about 100 nS for a UJT.

Table 5.1.4. *Programmable unijunction transistor definitions.*

$$R_{(min)} = 2(V_1 - Vv)1v$$

To obtain the minimum frequency use high values of capacitance, like 10 uF. Select devices and biasing for a low IP. Select low leakage capacitors such as those having glass or mylar dielectrics. The maximum timing resistor is found from:

$$R_{(max)} = (V_1 - Vp)21p$$

With allowance for capacitance and biasing variations, the approximate PUT frequency range is .003 Hz to 2.5 KHz.

5.1.5. The Silicon Controlled Rectifier (SCR, TRIAC)

The term *thyristor* may not be well known to us, but the devices are in our service whenever we use a light dimmer or motor speed controller, in operation of many appliances, and uses that might never enter our imaginations, such as control of the outdoor lighting at shopping centers and stadiums. Modern thyristors have spectacular power handling capacity.

Just when the term *thyristor* came into use is unknown to me. General Electric and Motorola were two corporations active in the early development of these devices. I have Motorola's *SCR Manual Including Triacs and Other Thyristors, Fifth Edition, 1972-Fifteenth SCR Anniversary*. This is a very comprehensive, informative book, revealing the longtime history of this solid-state device family. **Table 5.1.5** contains a listing of thyristor terminology.

The current *Motorola Thyristor Device Databook*[1] defines the thyristor "as solid-state switches which act as open circuits capable of withstanding the rated voltage until triggered." When triggered, these become low-resistance current conductors which continue until the current is reduced to its minimum holding value.

di/dt	Current rate of change. Generally applied as the maximum the device will accept without breakdown.
$I_T(RMS)$	Forward Current RMS. The maximum on-state current conduction.
I_{GM}, I_{GFM}	Forward Peak Gate Current. Maximum allowable gate current.
I_{TSM}	Peak Forward Surge Current. Maximum allowable non-repetitive surge current at a specified pulse width.
$I_T(AV)$	Average On-State Current. Maximum average on-state current the device can conduct with stated conditions.
P_{GM}	Peak Gate Power. Maximum instantaneous gate power dissipation between gate and cathode terminals.
$P_G(AV)$	Forward Average Gate Power. Maximum gate power averaged over a full cycle dissipated between gate and cathode.
I^2t	Circuit Fusing Considerations. Maximum forward non-repetitive overcurrent capability. Specified for one-half 60 Hz operation.
V_{GM}	Peak Gate Voltage. Maximum peak voltage between gate and cathode for any bias condition.
V_{FGM}, V_{GFM}	Peak Gate Voltage Forward. Maximum peak voltage allowed between the gate and cathode when forward biased.

1. *Motorola Incorporated, Thyristor Device Databook, DL137 REV 1, 1988.*

V_{RGM}, V_{GRM}	Peak Gate Voltage Reverse. Maximum peak voltage allowed between the gate and cathode when reverse biased.
V_{DRM}	Peak Repetitive Forward Blocking Voltage (SCR). Maximum repetitive forward voltage which may be applied without switching on the SCR.
V_{RRM}	Peak Repetitive Reverse Blocking Voltage (SCR). Maximum repetitive reverse voltage which may be applied to the anode terminal.
V_{DRM}	Peak Repetitive Forward Blocking Voltage (TRIAC). Maximum repetitive off-state voltage which may be applied without switching the triac.
I_{DRM}	Peak Forward Blocking Current (SCR). Maximum current which will flow at V_{DRM} and specified temperature.
I_{RRM}	Peak Forward Blocking Current (SCR). Maximum current which will flow at V_{RRM} and specified temperature.
I_{DRM}	Peak Forward Blocking Current (TRIAC). Maximum current which will flow for either polarity of V_{DRM} and at specified temperature.
V_{TM}	Peak On-State Voltage. Maximum voltage drop across the terminals under stated conditions.
I_{GT}	Gate Trigger Current. Maximum gate current required to switch the device from off- to on-states at specified conditions.
V_{GT}	Gate Trigger Voltage. Gate DC voltage required to produce the gate trigger current.
I_H	Holding Current. The forward anode current at which the device remains in conduction. Below this level the device reverts to the forward blocking state.
dv/dt	Critical Rise of Off-State Voltage. Minimum rate of rise of forward voltage resulting in off-state to on-state switching.
tgt	Turn-On Time (SCR). The time delay between the initiation of the gate pulse and the instant when device current switches to a specified on-state value.
tg	Turn-Off Time (SCR). The time delay between the instant when the anode current has decreased to zero following switching of the external voltage and the instant when the device will support a specified waveform without reverting to the on-state.

Table 5.1.5*. Thyristor terminology.*

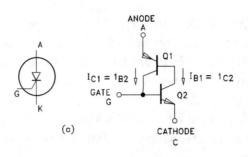

Figure 5.1.19. (a) The SCR electrical symbol. (b) The transistor circuit equivalent.

The Silicon Controlled Rectifier (SCR)

Some of us who have been around for a while recollect the *thyratron,* a gas filled tube used where a high energy switch was required. The gas tube had an anode connected to a high positive voltage, a cathode at or near ground, and a third terminal, called a grid. The grid consisted of a fine wire mesh positioned between the anode and the cathode to form an electrostatic screen. By keeping the grid at a negative potential relative to the cathode the tube's current conduction could be maintained in a standoff status. Once the grid went slightly positive relative to the cathode its control was lost; the only way to stop the current was to drop the anode potential below some minimum holding value.

In its operation, the silicon controlled rectifier (SCR) imitates the thyratron. Of course, that is where the resemblance ends.

My early experience with the SCR was the requirement for a very high energy pulse for the excitation of a fast high-pressure gas measurement gauge. Where significant power levels are involved the switch time must be short as possible to limit the internal dissipation as the device passes through its on/off transitions. In that instance I charged a capacitor to four hundred volts, then triggered an SCR to dump the charge into the gauge.

The easiest way to understand the action taking place in the SCR is with the two-transistor analogy of *Figure 5.1.19.* As an initial condition, assume the lead shown as *cathode* is grounded. The *gate* lead connects to ground

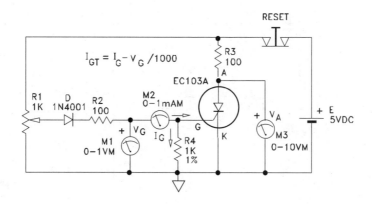

Figure 5.1.20. *A circuit for observing the DC transitions.*

through a small isolating resistance. Now a positive voltage is applied to the lead identified as *anode.* In this circumstance transistor Q2 is cutoff thereby preventing the flow of base current from Q1. Thus we observe that both devices are in the off state.

Now we raise the potential of the gate lead until just a smidgen of base-2 current flows. The reaction is swift as the regenerative feedback induces a flow of collector current from Q1 into the base of Q2 setting off a most impressive rush of current, driving both transistors into saturation. Now when we disconnect the external voltage to base-2 the transistor's status remains unchanged. Indeed, the only way to turn off the current is to reduce the anode current to below its holding value, very nearly zero, at which time the anode potential swiftly returns to the supply level.

We can take a measure of the action by breadboarding the circuit of *Figure 5.1.20*[2], with the external components and meters shown. With my circuit the switching took place with the gate at 0.8 volts. The gate current was 1.05 mA. The anode voltage dropped from 5.00 volts to 0.75 volts.

Like any semiconductor device, the SCR can be destroyed by excessive heat dissipation arising from usage beyond its ratings and/or inadequate heat sinking. However, there is another source of failure unique to these

2. *Teccor Electronics, Inc., A Siebe Company, General Catalog,*

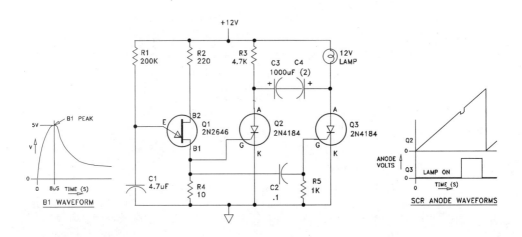

Figure 5.1.21. A commutating lamp flasher with UJT B1 and SCR anode waveforms.

devices, called the *di/dt effect. di/dt* is a mathematical term defining the rate of change of current with time. As used here it relates to the SCR's turn-on mechanism. When bias current is forced into the gate, a rapid internal up-heaval takes place as the anode-to-cathode current rises accompanied by a swift drop in voltage across the device. Depending on the SCR type the transition time may be anywhere from a few nanoseconds to several micro-seconds.

The dissipation internally within any switching device is greatest during its transitions: off-to-on and on-to-off. There is some point in the conduction when the product of junction current and voltage is at a peak. Frequently the current is initially channeled through a narrow region of the junction, which then must assume the entire burden of the heat dissipation. The ultimate limit is the junction temperature. For reliable operation, the tem-perature must always be maintained below the manufacturer's defined maximum.

Failure is most likely to occur at turn on. Two means of avoiding di/dt mode failure are: (1) increase the gate drive, and (2) limit the anode current rise time. When a small gate drive is used, the conduction is initiated at a small spot size in the junction. Increasing the drive spreads this region out quicker. A small inductance in the anode circuit slows down the current rise time.

Figure 5.1.21 is the circuit for a lamp flasher[3]. The flashing is effected by commutating the anode current of two SCRs, much like the discrete monostable. The value of the components shown differ in some respects from those in the referenced circuit. They worked for the 2N4184 SCR after some playing around. Note the notch in the anode voltage of the first SCR. The turnoff of Q3 did not switch off Q2 though an attempt was made, but its turn-on did. Some additional experimentation would most likely remove this effect and more closely approach an equal time division.

The TRIAC

Since its introduction in the late 1950s the SCR has grown into a large family of devices capable of switching extremely large currents, both DC and AC. Available products include:

Triacs:
    Bidirectional AC switches available with load range capabilities of 1 - 40 RMS amperes. Used for the control of full-wave AC power either through full-cycle switching or phase control of the load current. Devices capable of blocking up to 800 volts are available. Typical applications include lighting, heating, and motor speed control.

Logic Triacs:
    AC bidirectional switches that respond to single polarity switching.

Quadracs:
    A *triac* with a built-in *diac* trigger.

Alternistors:
    Two electrically separate SCR structures providing enhanced di/dt characteristics.

Diacs and Silicon Trigger Switches (STS):
    Trigger devices used in phase control circuits to provide gate pulses to an SCR or triac. They are voltage triggered bidirectional devices in a small glass axial lead package.

3. Reference D. R. Grafham, General Electric Company, Using Low Current SCRs, Applications Engineering, Syracuse, New York, 200.19 1/67.

Sidacs:

> A bidirectional voltage-triggered switch. Features include a one-cycle surge capability up to 20 amperes, 95 - 330 volt switching range, turn-on latching, and low on-state resistance. Ideal for dumping charged capacitors through an inductor for the generation of high power pulses. (Too bad they didn't exist back in 1970 when I needed one.)

The triac symbol and voltage-current characteristics are described in **Figure 5.1.22**. Note how the current and voltage both pass through zero in their transitions between quadrants. Re-triggering is required to maintain conduction. The logic triac features both AC and DC triggering. With AC or positive unipolar pulse triggering quadrants I and III are used. With negative unipolar pulse triggering quadrants II and IV are used. **Figure 5.1.23** defines quadrant assignments.

Triac Triggers

Figure 5.1.24 illustrates the functioning of a *silicon bilateral switch (SBS)* as a triac trigger source. The SBS is a semiconductor with negative resistance switching similar to the 4-layer diode and unijunction transistor as well as the *diac* and *STS*.

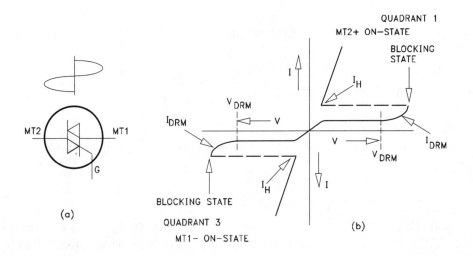

Figure 5.1.22. (a) Triac symbol. (b) Triac voltage-current relationships.

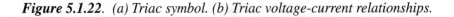

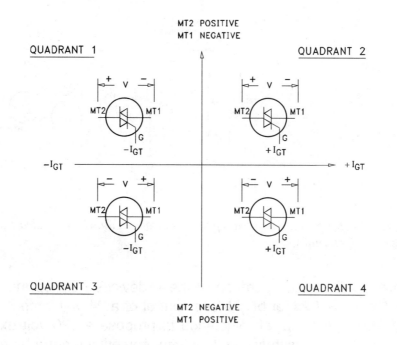

Figure 5.1.23. *Quadrant definitions.*

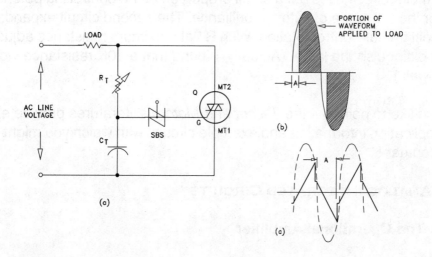

Figure 5.1.24. *(a) A basic phase controlled triac circuit. (b) Phase controlled AC wave-form. (c) Capacitor voltage waveform.*

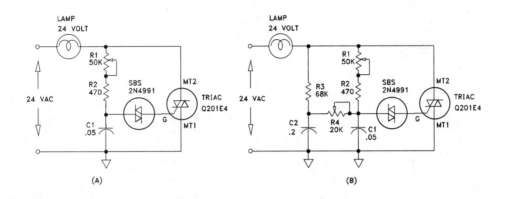

Figure 5.1.25. (a) Basic full-wave triac phase control circuit. (b) Extended range full-wave phase control circuit.

To obtain a "feel" for the operation of these devices I set up the circuits shown in **Figure 5.1.25** for brightness control of a 24 volt lamp. I used a stepdown transformer out of reluctance to propose a 120 volt example. Those of us who work with five volt circuitry sometimes get a bit relaxed. 120 volts is dangerous. If you are new to these devices, keep in mind the peak-to-peak voltage is close to 200 when it comes to capacitor ratings.

The simpler circuit uses but a small proportion of the controlling potentiometer for the full range of cutoff to brilliance. The second circuit expands this to almost the full control rotation. With R1 at maximum resistance adjust R4 to just extinguish the lamp. (Actually, I found that a 10K resistance worked pretty well.)

I should like to point out the *Teccor* and *Motorola* literatures provide excellent application information and example circuits with which you might wish to experiment.

5.2. ANALOG INTEGRATED CIRCUITS

5.2.1. The Operational Amplifier

In 1956, George A. Philbrick Researches, Inc., published its *Application Manual For PHILBRICK Octal Plug-in Computing Amplifiers.* About a quar ter of a million copies through ten editions were distributed. Ten years later, a follow-up publication, *Application Manual for Computing Amplifiers for Modeling Measuring Manipulating & Much Else*[1], made its debut. If you can find this somewhere, it is a real treasure of analog's "who, what, why, how, and when." In the 1950s and '60s, and for some time beyond, analog computers were widely employed as research and development tools. The ubiquitous op amp made it possible.

The analog integrated circuit (IC) briefly trailed its digital predecessors; but by 1966 solid-state and vacuum tube op amps were about equally in use. The early IC op amps were not always that easy to use. Amplifier circuits frequently made excellent oscillators. Those that didn't were frequent high-end duds. All the early guys required compensation and offset adjustment. Slew rates weren't, and a phenomenon called "popcorn noise" received a lot of press. One of the earliest, Fairchild Corporation's mA741 has proved to be exceptionally durable. I still have a couple of Amelco 809s from the 1960s; a pretty good op amp for its time. My 1989 copy of the National Semiconductor Corporation's *General Purpose Linear Devices Databook* includes 816 pages just for the op amp — and that is only part of it.

Ah well, so much for nostalgia.

There are far more possible applications of the op amp than can ever be discussed here. I strongly recommend a bit of research in the manufacturers' data books and the great variety of technical literature now available[2,3,4,5]. While not an in-depth presentation, what we find here will give us a basic understanding of the op amp and some of its uses.

Figure 5.2.1 illustrates op amp circuitry for the majority of tasks to which it is suited. The basic non-inverting (b) and inverting (c) configurations are

1. *George A. Philbrick Researches, Inc., Applications Manual For Computing Amplifiers For Modeling Measuring Manipulating & Much Else, 1966.*

2. *Thomas M. Frederiksen, Intuitive IC Op Amps, Linear IC Design Group, National Semiconductor Corporation, Santa Clara, CA 1984.*

3. *David F. Stout and Milton Kaufman, Editor, Handbook of Operational Amplifier Circuit Design, McGraw-Hill Book Co., 1976.*

4. *Sol D. Prensky, Manual of Linear Integrated Circuits, Reston Publishing Co., Inc., Reston, VA, 1974.*

5. *Jerald G. Graeme, Gene E. Tobey, Lawrence P. Huelsman, Operational Amplifiers - Design and Applications, McGraw-Hill Book Co., 1971.*

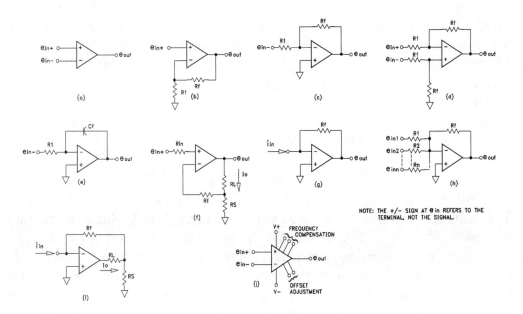

Figure 5.2.1. *(a) Conventional operational amplifier (op amp) symbol. (b) Noninverting amplifier. (c) Inverting amplifier. (d) Differential amplifier. (e) Op amp integrator. (f) Voltage-to-current converter. (g) current-to-voltage converter. (h) Summing amplifier. (i) Current-to-current converter. (j) Other possible terminal provisions.*

most widely used, and are embedded to some extent in those that follow. If we eliminate resistor R1 and short out Rf in circuit (b) we have the popular unity gain voltage follower. (j) shows the external connections we may find on our op amp depending on its type. Note the absence of a ground terminal. Early types did have one, but the new ones, that I am aware of and use at least, do not. So where does the op amp find its reference to ground? Where else but in the external circuit connections.

Figure 5.2.2 shows six of the more popular applications with equations for the ideal amplifier. For the majority of projects we hobbyist/experimenters get involved with the ideal assumption is sufficient. For this to be true, however, requires care in our choice of amplifiers.

Table 5.2.1 lists the major op amp parameters. **Table 5.2.2** lists the most significant design equations. While some variation between manufacturers exists, for the most part the terminology is standardized.

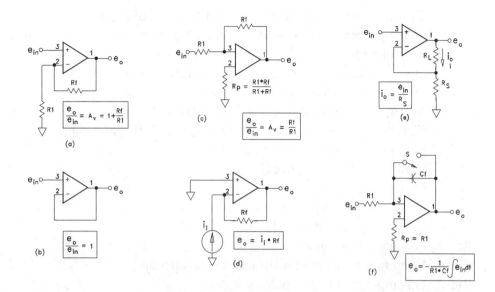

Figure 5.2.2. *Some common applications of the operational amplifier. (a) Noninverting amplifier circuit. (b) Noninverting follower circuit. (c) Inverting amplifier circuit. (d) Current-to-voltage (transresistance) amplifier circuit. (e) Voltage-to-current (transconductance) amplifier circuit. (f) Integrating amplifier circuit.*

Do not neglect the manufacturers' text and graphic performance characteristics provided in the databooks. This is where the essential details of input/output voltage and current, small and large signal and frequency response and slew rates and noise currents and voltages are all spelled out, along with a plethora of application examples.

A significant factor in our amplifier selection is the *input bias current*, Ib. The infinite input resistance of the ideal amplifier draws no bias current. So with the inverting circuit all the incoming current is from the source and flows on through the feedback resistance to the output. (Some literature will show If flowing toward the input, preceded with a minus sign.)

Similarly for the amplifier output — the ideal features a zero internal resistance so all the gain appears across the load resistor, RL.

However, in the amplifier's real world, there is input bias current and a finite internal resistance at the inputs and output. These are illustrated in the (b)

Av	Op amp open loop voltage gain as a function of frequency.
Avo	Op amp open loop DC voltage gain. May be substituted for Av in any equation when DC gain is the objective.
Avc	Op amp closed loop voltage gain as a function of frequency.
Avco	Op amp closed loop DC voltage gain.
ß	Voltage feedback ratio. ß = Ri/(Ri + Rf)
fcp	First pole frequency of the circuit, i.e., the -3 dB bandwidth.
fop	First pole frequency of the op amp.
Ib	Op amp input bias current.
Iio	Op amp input offset current.
In	Op amp input equivalent noise current.
Ri	Input resistance, external to the op amp.
Rf	Feedback resistance.
Ric	Common-mode input resistance of the op amp.
Rid	Op amp differential input resistance.
Rin	The circuit input resistance.
RL	The circuit load resistance.
Ro	The op amp source resistance, i.e., looking back in.
Rout	The circuit's output resistance.
tr	The circuit's risetime, 10 to 90%.
Vio	Op amp input offset
Vn	Op amp input equivalent noise voltage.
Von	The circuit's output noise voltage.
Vo	The circuit's output voltage.

Table 5.2.1. Op amp design parameters.

$Avc = -Rf/R1$ The op amp ideal closed loop voltage gain.

$Avc = -Rf \cdot R1/(1 + 1/ßAv)$

The op amp closed loop voltage gain with the finite op amp gain included. $ß = R1/(R1 + R2)$

$Avc = -Rf \cdot R1/(1 + 1/ßAv + 2Rf/AvRid)$

The op amp closed loop voltage gain with the differential input resistance and op amp gain included.

Avc = -Rf • R1/(1 + (Rf + Ro)/ßAv • Rf)

> The op amp closed loop voltage gain with the op amp output resistance and op amp gain included.

Rin = R1   The circuit input resistance assuming ideal op amp parameters.

Rin = R1(1 + Rf/Avo • R1)

> The circuit input resistance assuming finite Avo.

Rout = 0   The circuit output resistance assuming ideal op amp parameters.

Rout = Ro(1 + /ß • Av)

> The circuit output resistance assuming op amp finite output resistance and Avo.

fcp = fop • Avo • R1/Rf

> Circuit bandwidth assuming the -3db op amp bandwidth is at fop, the first pole of the op amp.

tr = 0.35 • Rf/fop • Avo • R1

> Circuit small signal rise time, 10-90%.

_Vo = ±_Vio(R1 + Rf)/R1

> Incremental output DC voltage change arising from input offset voltage change, assuming Ib and Iio = 0.

Vo = Ib • Rf   Output voltage change due to input bias current, assuming Rp = 0 and Vio = 0.

_Vo = ±_Iio • Rf

> Incremental output DC voltage change arising from input offset current change, assuming Rp = R1 • Rf/(R1 + Rf) and Vio = 0.

Von = Vn(1 + Rf/R1) • V/Hz

> Output noise voltage arising from an equivalent op amp input noise voltage in volts/Hz.

Von = Vn[V²n(1 + Rf/R1)² + In² • Rn²]exp½

> Output noise voltage arising from both an equivalent op amp input noise voltage and current noise, V/Hz and A/Hz or V²/Hz and A²/Hz.

Rp = R1 • Rf/(R1 + Rf)

> The optimum value for Rp to minimize output voltage offset arising from Ib.

Table 5.2.2. Op amp design equations.

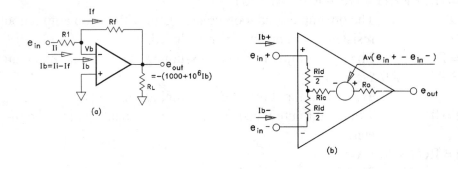

Figure 5.2.3. *Op Amp input biasing. (a) Basic inverting amplifier. (b) Definition of amplifier input and output stage equivalent resistances and voltage gains.*

portion of **Figure 5.2.3**. So, voltage division takes place between our circuit's source resistance, R1, and the amplifier input. If our source resistance is large and the amplifier's input small, we could have a problem. In a similar manner we want to tailor the load resistance so as to minimize the internal amplifier losses. This has merit beyond conserving voltage gain — internal losses means heat generation, which further exacerbates the drift problems that can arise.

When the input bias is sufficient to induce a noticeable offset it must be dwelt with. This problem typically arises when we have a low level signal which requires a lot of voltage gain. The two bias currents are never exactly equal and while their temperature drifts will be close the voltage drops will vary even so. **Figure 5.2.4** illustrates three possible solutions, depending on the amplifier and the circuit configuration. (a) of the figure shows the connections for the LF355/6/7 series of FET input amplifiers.

This same amplifier is shown in **Figure 5.2.5**, where its input stage circuitry is compared to that of a bipolar amplifier, here the LM776. The very high input resistance of the FET amplifiers is frequently helpful in the solution of the input drift with temperature problem. MOS/CMOS inputs are susceptible to damage from static electricity and must be handled with care.

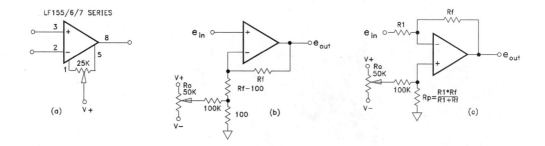

Figure 5.2.4. *Some approaches to input offset compensation. (a) Amplifier internal provision (LF355/6/7 example). (b) External compensation, inverting amplifier. (c) External compensation, noninverting amplifier.*

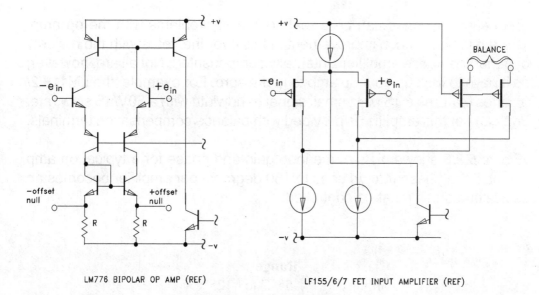

Figure 5.2.5. *Op Amp Input comparisons: (a) A typical bipolar amplifier circuit (the LM776). (b) FET amplifier circuit (LM355/6/7).*

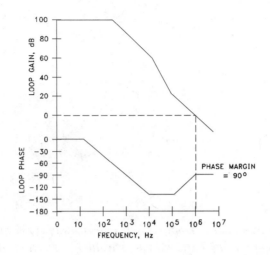

Figure 5.2.6. Loop gain and loop-phase plots for a typical op amp circuit.

Frequency response can be a real source of problems with the op amp. There has been much improvement in this over the years with the majority of general purpose amplifiers internally compensated. Not all are, however; so it pays to look up your op amp to make sure. For example, the LM118/2/3 series features a 15 MHz small signal bandwidth with a 50V/µS slew rate. You can bet this amplifier is provided with balance/compensation terminals.

Figure 5.2.6 shows plots of the loop gain and phase for a typical op amp circuit. If the phase margin goes to 180 degrees, our amplifier becomes an oscillator, so this must be controlled.

Class	Range
Military	-55°C to +125°C
Automotive	-40°C to +85°C
Industrial	-25°C to +85°C
Commercial	0°C to +70°C

Table 5.2.3. Manufacturers' temperature range classifications.

DEVICE	#FEATURES	VOS	IBIAS	PACKAGE	PRICE
LM10CLN	Inc.Pre.Ref.	2mV	≤20nA	8-Dip	$5.25
LM10CWM	Inc.Pre.Ref.	2mV	≤20nA	SO-14	$8.75
LM301AN	Gen. Purpose	7.5mV	≤250nA	8-Dip	$1.12
LM307AN	Compen LM301A	7.5mV	≤7nA	8-Dip	$1.40
LM308AN	Imprvd LM301A	7.5mV	≤7nA	8-Dip	$1.75
LM308AM	Imprvd LM301A	7.5mV	≤7nA	SO-8	$1.30
LM318N	Hi Slew Rate	4mV	≤500nA	8-Dip	$1.68
LM318M	Hi Slew Rate	4mV	≤500nA	SO-8	$3.22
LM324AN	Lo Pwr Quad	7mV	≤250nA	14-Dip	$1.05
LM324AM	Lo Pwr Quad	7mV	≤250nA	SO-14	$0.88
LM741CN	Gen. Purpose	6mV	≤500nA	8-Dip	$0.91
LM741H	Hi Prf,Hi Tmp	6mV	≤500nA	TO-99	$4.90
LM741CM	Gen. Purpose	6mV	≤500nA	SO-8	$0.91
LF351N	Bi-Fet Gen Pur	10mV	≤200nA	8-Dip	$0.75
LF351M	Bi-Fet Gen Pur	10mV	≤200nA	8-Dip	$0.75
LF353N	Dual Bi-Fet	10mV	≤200nA	8-Dip	$0.75
LF353M	Dual Bi-Fet	10mV	≤200nA	SO-8	$0.75
LF347N	Quad Bi-Fet	5mV	≤100pA	14-Dip	$1.60
LF347M	Quad Bi-Fet	5mV	≤100pA	SO-14	$1.60
LF355N	Bi-Fet Lo Pwr	10mV	≤200nA	8-Dip	$0.75
LF355M	Bi-Fet Lo Pwr	10mV	≤200nA	SO-8	$0.75
LF356N	Bi-Fet Wd Bnd	10mV	≤200nA	8-Dip	$1.35
LF356M	Bi-Fet Wd Bnd	10mV	≤200nA	SO-8	$1.50
LM2/3900	Quad, S/D Sup	——	≤200nA	14-Dip	$2.83
LT1001CN	Lo Ofst, Drft	60mV	≤2nA	8-Dip	$2.88
LT1006CN	5V Sup,Hi Per	50mV	≤15nA	8-Dip	$2.88
LT1055CN	HiPer JFETinp	400mV	≤50pA	8-Dip	$3.38
TLC252AC	LinCMOS,Hi Ib	2mM	≤1pA	8-Dip	——
TL287C	BiFET Gen Pur	.5mV	≤200pA	8-Dip	——

Table 5.2.4. Some representative commercial bipolar and JFET input amplifiers. The prices shown may not be current and are for comparison purposes only.

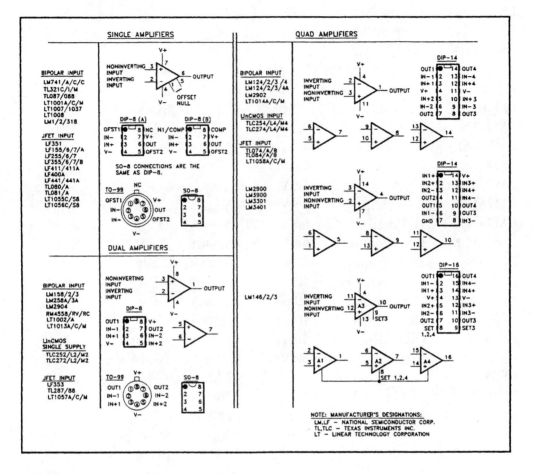

Figure 5.2.7. *Some amplifier types with model numbers, symbols, and case styles.*

Most op amps require dual power supplies, being unable to fully function with their inputs and outputs close to ground. There are some that are tailored for single power operation referenced to ground: examples are the LM324 and *Linear Technology* LT1006.

There are literally hundreds of excellent op amps available — so many that only a representative few can be described here. ***Table 5.2.3*** defines the four temperature classifications employed by manufacturers. ***Table 5.2.4*** lists a variety of commercial range amplifiers with some parameters. The prices listed are contemporary, but shown here only for comparison within the table. *Figure 5.2.7* illustrates the electrical symbol and package style for a variety of available op amps. While the tables and figure can be helpful in amplifier understanding and selection, they are no substitute for the manufacturers' data.

5.2.2. THE ANALOG COMPARATOR

We use the comparator to convert a changing analog signal into two distinct limit values. Not too surprisingly, the analog comparator exhibits a very close relationship with the operational amplifier. Internally they are very much the same; the differences arise in the enhancement of open loop gain, frequency response, slew rate, and output characteristics. With the op amp in its usual applications, we look for gain linearity with frequency over a specified output swing. Our frequent need with the comparator is for fast, level sensitive, repeatable response and output compatibility to one or more logic families. In most respects we can view the comparator performance in op amp terms other than the option for hysteresis.

There are eight distinct comparator circuits we can consider[1]. They are:

1. An inverting zero-crossing detector.
2. A non-inverting zero-crossing detector.
3. An inverting zero-crossing detector with hysteresis.
4. A non-inverting zero-crossing detector with hysteresis.
5. An inverting level detector.
6. A non-inverting level detector.
7. An inverting level detector with hysteresis.
8. A non-inverting level detector with hysteresis.

First let's agree on the definition of hysteresis, which is formally: "The failure of a property that has been changed by an external agent to return to its original value when the cause of the change has been removed." Basically it is a stabilizing element; without it our circuit may very well go into a state of the jitters, careening uncontrollably between two limits.

1. *David F. Stout and Milton Kaufman, Editor, Handbook of Operational Amplifier Circuit Design, McGraw-Hill Book Co., 1976*

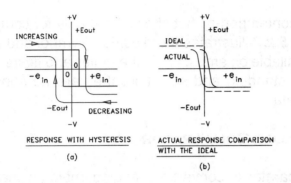

Figure 5.2.8. (a) Zero detection with hysteresis. (b) The actual response vs the ideal.

Figure 5.2.8(a) is a graphical representation of hysteresis. The effect is produced by a small amount of positive feedback that must be overcome to initiate the return transition. The (b) portion relates the voltage-transfer of the ideal amplifier to the actual. What we observe is a reduced output level due to internal impedances under external circuit loading and reduced rise and fall times. The latter results from amplifier slew rate and gain limitations in combination with external effects.

Note, however, the comparator is not responding to the rise or fall time of the input: it is level-sensitive only. In the absence of hysteresis, undesired effects around the transition through zero may occur — in particular with slow input changes or signal noise undesired output switching may occur. This is aggravated by a high-gain, high-speed amplifier.

Amplifier slew rate places an upper limit on speed of the output transition. External circuit impedances are also influential. Circuit stability comes with a price: careful evaluation of the requirements; the selection of an appropriate amplifier (comparator); and suitable input and output circuitry. With hysteresis as needed, too little may result in random erratic behavior; too much will drag down the response unnecessarily.

Figure 5.2.9 gives us a look at three illustrative circuits. In these, the comparator output is assumed to be an open transistor collector or its equivalent in terms of performance; that is, the limits are set by the character of the load. Pull-up resistors, R_{pu}, and zener diode clamps, Vz, are commonly used.

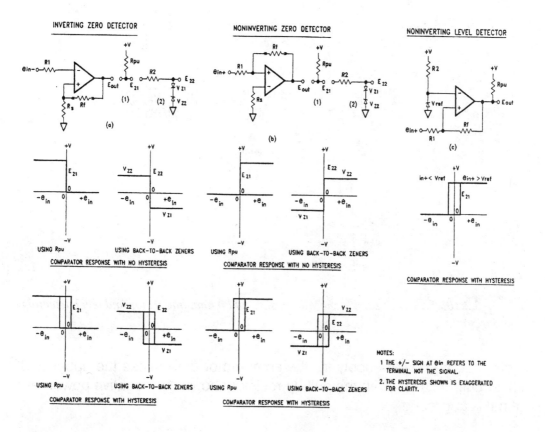

Figure 5.2.9. Defining inverting and noninverting zero detection and a mode of level detection. (a) Inverting zero. (b) Noninverting zero. (c) Noninverting level.

An Inverting Zero-Crossing Detector

Shown in *Figure 5.2.9*(a), this circuit determines the status of the voltage across its terminals relative to a zero reference, typically ground. We are looking for a response at the output to tell us whether the signal is greater than or less than zero. We do not want to be told it is exactly at zero. In this circuit a positive response is obtained from a negative (-) input transition. The graphs show an ideal response, with and without hysteresis. In this portrayal the feedback resistor, Rf, is shown with brackets indicating it may or may not be present.

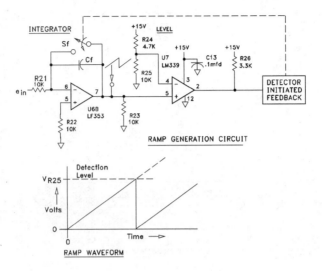

Figure 5.2.10. *Sawtooth ramp generation using an op amp integrator and level detecting comparator.*

Note that the power supply for the comparator determines the upper and lower output limits. With the pull-up resistor shown the negative power terminal is at ground.

A Noninverting Zero-Crossing Detector

Shown in **Figure 5.2.9**(b), this circuit is a reversal of the inverting, zero-crossing detector, providing a positive response to a positive, (+), input transition. Other than for this the behavior of the two configurations is the same.

A major advantage in using the non-inverting circuit is the reduced loading on the source.

A Non-Inverting Level Sensing Detector

Shown in **Figure 5.2.9**(c), the comparator's inverting terminal is connected to a source of stable voltage. The voltage at the noninverting terminal must exceed this reference to effect an output transition. The presence of the

Part No.	Response Time (typ) nS	VOS mV(Max)	IS mA(Max)	IB nA(Max)	Descript.
LM361	12	5	25	30,000	High speed w/strobes
LM360	16	5	32	20,000	Hi spd w/ comp.outputs
LM306	40	5	10	25,000	High speed, high drive
LM319	80	8	12.5	1,000	High speed dual
LF311	200	10	7.5	0.15	FET input single
LM311	200	10	7.5	300	General purpose single
LH2311	200	7.5	7.5	250	Dual LM311
LM339	1300	5	2	400	General purpose quad
LM392	1300	10	1	400	1 comp w/1 op amp
LM393	1300	5	2.5	250	General purpose dual
LM2903	1300	5	2.5	250	Automotive dual
LM2901	1300	7	2	400	Automotive quad
LM613	1500	5	1	35	Dual comp, dual op amp Adjustable reference
LP365	4000	9	0.30	200	Program-mable quad
LP311	4000	10	0.3	150	Low power single
LP339	5000	9	0.1	40	Low power quad

Table 5.2.5. Voltage comparator selection guide, commercial grade devices.

feedback resistor means this input must drop slightly, in practice very slightly, below the reference to bring about the return transition. Too much positive feedback will delay the transition or even bring on oscillations.

A Functional Sawtooth Ramp

Figure 5.2.10 shows the circuit in use for creating the ramp in the waveform generator of Chapter 3, **Figure 3.4.1**. In this application, an op amp integrator provides the ramp. The switch in parallel with the feedback capacitor is periodically closed to discharge the capacitor, thus reinitializing the sequence. Ramp length in this circuit is established at the input to resistor R21. The ramp height is a constant determined by the voltage drop across R25. Varying the input to the op amp determines the charging time of the feedback capacitor and hence the length of the sawtooth. In the working circuit the *detector initiated feedback* consists of monostables and analog switches. Note the absence of hysteresis. The comparator triggers a monostable and the reset action goes forth from there.

Design Procedure

The first requirement is to acquire a good understanding of the objective.

Exactly what will be required of the circuit? Is the comparator to control an analog or a digital device? This most likely will influence the output swing - between two voltage limits or a single voltage and ground.

What is the speed requirement? We need an understanding of the minimum slew rate, measured in volts/microsecond (V/mS), as this may impose limitations on the comparator we can use and its circuit interactions.

Is there a concern for stability? If so, then how much hysteresis should be provided? This is often worked out with a realistic breadboard at the bench.

Table 5.2.5 is a selection guide of comparators with the commercial temperature classification.

5.2.3. The Waveform Generator

The ICL8038 Precision Waveform Generator/Voltage Controlled Oscillator (VCO)

Sooner or later, a good sine wave generator becomes a must-have tool for our home lab. The ICL8038[1] is an economical, easy-to-use monolithic IC well suited to our needs. A functioning generator can be up and running with the addition of a few resistors, trimpots, and capacitors. Input and output op amp buffers extend its interfacing flexibility.

This IC is able to produce precision sine, triangle, and square wave outputs over a frequency range 0.001 Hz to 300 KHz. Other features as described in the Harris databook for linear ICs are[2]:

- Low frequency drift with temperature - 250 ppm/°C.
- Low 1% sine wave distortion.
- High triangle wave .01% linearity.
- Duty cycle variable over the range of 2% to 98%.
- TTL compatible square wave output.
- Sweeping and frequency modulation capability.
- May be interfaced with phase locked loop circuitry.

Figure 5.2.11 is a function diagram showing the internal operation of the IC. A pin diagram is also shown. The operation is based on two current sources, labelled I and 2I, which alternately charge and discharge an external capacitor. Source #1 furnishes a continuous charging current. Source #2 outputs a subtracting current double that of source #1. Two comparators track the capacitor voltage. Their response triggers a flip-flop for controlling the switching of current source #2. With the currents as defined the capacitor is charged and discharged with a net current I resulting in a symmetrical voltage that rises and falls linearly with time.

In addition to switch control, the flip-flop provides a buffered square wave. The buffer is an uncommitted NPN transistor collector. This allows us to

---

1. *For further information on the ICL8038 I suggest the Harris Semiconductor Corporation Application Note A013, Everything You Always Wanted To Know About the ICL8038.*
2. *Harris Semiconductor Corporation, Linear ICs For Commercial Applications, 1990.*

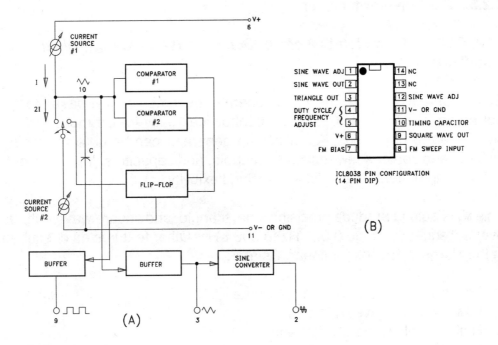

Figure 5.2.11. (a) The ICL8038 function diagram. (b) ICL8038 pin assignments.

employ an appropriate pull-up resistor for voltage appropriate to the next stage: +5V for TTL logic, for instance.

Figure 5.2.12 is the basic circuit for a sine wave generator. Resistors RA and RB are the emitter resistors for internal current sources. When these are equal a symmetrical sine wave with minimal distortion results. Resistor RA controls the rising side of the triangle and sine waves and the 1 (High) state of the square. The magnitude of the triangle waveform is set at 1/3 of the power supply which leads to the following definition for the rise time:

$$t1 = C \bullet V/I = (C/3) \bullet V \bullet RA/0.22 \bullet V = RA \bullet C/0.66$$

The **V** term here refers to the total power supply voltage. For a dual supply, that is the sum of the plus and minus values.

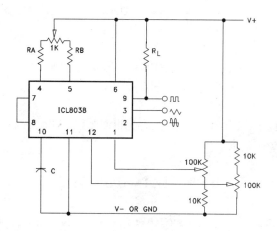

Figure 5.2.12. The ICL8038 optimum distortion circuit.

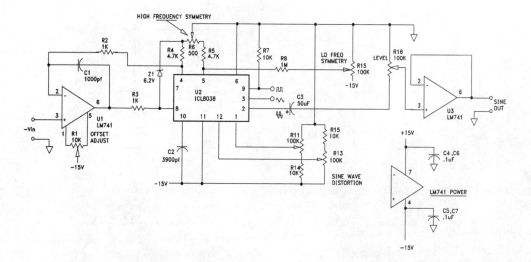

Figure 5.2.13. A sine wave circuit with a 20Hz to 20KHz frequency range.

The falling portion of the two waves and the 0 (low) state of the square wave is:

$$t2 = C \bullet V/I = C/3 \bullet V \bullet RA/2 \bullet (0.22) \bullet V/RB - 0.22 \bullet V/RA = RA \bullet RB \bullet C/0.66(2RA - RB)$$

Note the power supply term, V, is cancelled out. When RA = RB, a 50% duty cycle is achieved. If:

$$RA = RB = R, f = 0.33/R \bullet C$$

For this condition, the 1K pot in *Figure 5.2.12* is assumed equally divided between the two resistances.

For any given frequency there are many combinations of RA, RB, and C that will do the job. There are, however, constraints on the range of the charging currents for the best performance. Currents less than one microamp enable leakage currents to contribute errors. For currents greater than 5

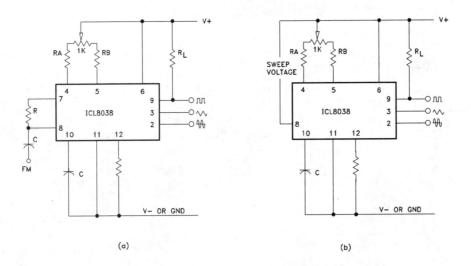

(a)   (b)

Figure 5.2.14. (a) An arrangement for FM modulation. (b) A sweep input circuit arrangement.

mA, transistor saturation voltages will induce errors. Best results are with currents between 10 µA and 1 mA. Then:

$$I = R1 \bullet (V(+) - V(-))/(R1 + R2) \bullet 1/RA = 0.22 \bullet (V(+) - V(-))/RA$$

— where R1 = 11 KΩ, R2 = 39 KΩ. A similar calculation applies to RB. The capacitor value is best when close to its upper end for the given range.

The power supply may be single, 10 to 30 volts, or dual, ±5V to ±15V.

An Application

Figure 5.2.13 is the schematic for a voltage controlled generator having a frequency range of 20 Hz to 20 KHz. The input amplifier buffers the controlling source while providing a low resistance source for the generator. The source resistance looking back into the sine wave terminal, pin 2, is on the order of 1 KΩ. The output amplifier provides a buffer, as well as a means of adjusting the output level. This is a circuit worthy of breadboarding to obtain an understanding of the device. And it could be constructed as a low cost, convenient module where it is not desirable to build the entire waveform generator of chapter three.

Frequency Modulation and Sweeping

The frequency is directly related to the DC voltage present at pin 8, measured from the positive supply rail. For a small deviation, on the order of 10%, an AC input can be capacitively coupled. This is shown in **Figure 5.2.14** (a). For a greater degree of modulation or sweeping the signal is applied between the positive rail and pin 8. This is shown in **Figure 5.2.14** (b). In this situation the supply voltage must be well regulated as any fluctuations will affect the triggering level. Pin 8's potential may be swept down from V+ by 1/3 • VSupply-2 • V. Note that the supply voltage reference is to the sum of the plus and minus power.

5.2.4. The Precision Voltage-to-Frequency Converter

The LM331/LM331A Precision Voltage-to-Frequency Converter

The LM331 is a real jewel when we need a simple approach to analog-to-digital conversion. This low-cost, easy-to-use integrated circuit plus a few resistors and capacitors is all that it takes for reliable, precision performance[1].

In addition to voltage-to-frequency conversion, the device has application for frequency-to-voltage conversion, long-term integration and linear frequency modulation/demodulation. The **A** version has improved temperature specifications. A band-gap reference circuit in the IC assures accurate conversion over the full specified temperature range at power supply voltages as low as four volts. The timer circuit incorporates low bias currents while not degrading the 100 KHz response. The frequency output terminal is an open collector NPN transistor. This enables pull-up to any logic voltage within the device maximum rating of 40 volts. Other features include:

- Guaranteed linearity of .01%.
- Split or single power supply operation, 5V£Vs£40V.
- Logic family compatible pulse output.
- Temperature stability - ±50ppm/°C max.
- Power dissipation of 15 mW at 5V typical.
- Full scale frequency range 1 Hz to 100 KHz.

Figure 5.2.15 is the simplified function diagram. The figure includes the pin assignments for the dual-in-line package. *Figure 5.4.16* is the schematic of the databook's standard test and application circuit. These two figures provide the insight required to adapt the IC to most conversion needs. However, as always, the manufacturer's data should be consulted for assurance that our circuit will perform as expected.

The cited databook includes many possible applications of the LM331 including frequency-to-voltage, light intensity-to-frequency, temperature-to-frequency, optically isolated conversion, and basic A/D conversion with computer interfacing.

1. *National Semiconductor Corporation, Data Acquisition Linear Devices Databook, 1989.*

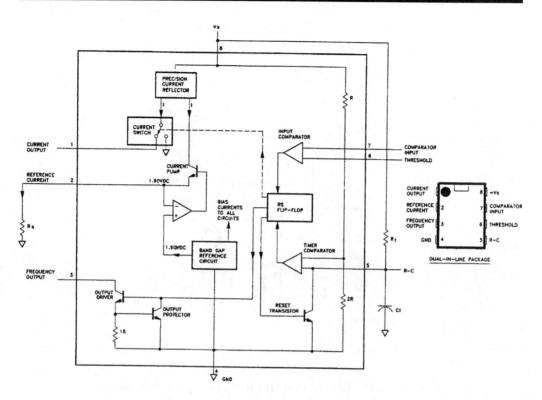

Figure 5.2.15. LM331 function diagram.

Figure 5.2.17 is the circuit for a frequency-to-voltage converter. This could be an interesting driver for an analog voltmeter calibrated to display engine RPM — or a 10-step LED level meter such as the Panasonic LN10204P. All manner of ideas come to mind!

5.2.5. The CMOS Analog Switch

The MC1\CD4016\MC1\CD4066 Quad Analog Switch\Multiplexer

These two devices are pin-for-pin replacements except for a much improved ON resistance, Ron, for the MC1\CD4066[1,2]. (Motorola devices employ the MC1 prefix, but are otherwise identical.) The switch has both analog and digital switching capability. An example of its use is resetting the integrator in the ramp generation circuitry of ***Figure 3.4.1.***

---

1. *Motorola Corporation, CMOS Logic Databook, DL131/D, Rev.3.*
2. *National Semiconductor Corporation, CMOS Logic Databook, Rev.1, 1988.*

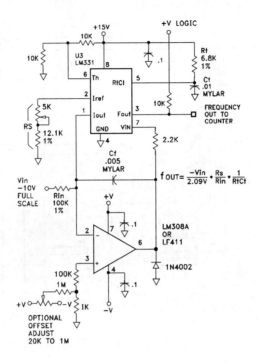

Figure 5.2.16. LM331 standard test and application circuit.

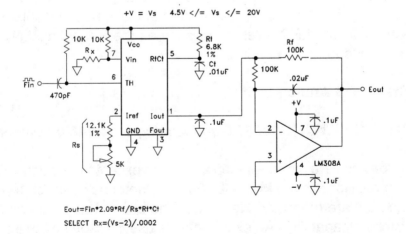

Figure 5.2.17. Frequency-to-voltage converter circuit. Full scale output is 10 volts, with 2-pole filter, ±.01%.

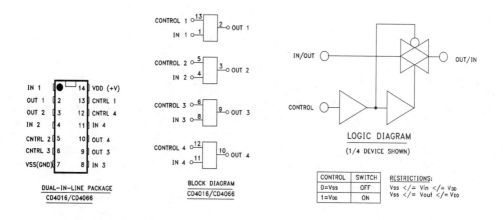

Figure 5.2.18. *The CD4016/CD4066 quad analog switch/multiplexer IC pin assignments, block diagram, logic diagram, and truth table.*

Devices of this type are named switch/multiplexer because they can be used as rotary switches in many applications. If the switch outputs connect to a common point, an audio system for instance, then any compatible source can be selected by simple input switching.

Figure 5.2.18 provides basic information on the device packaging, functions, logic symbol, and truth table. It is seen that the analog signal range is determined by the DC supply and control voltages; that is, the peak-to-peak AC signal must not exceed the DC voltage range, typically ground, V_{ss}, to the positive supply, V_{DD}. Protection against over/under range transients can be afforded by small signal protective diodes capable of absorbing possible current surges while clipping.

The MC1\CD4051/2/3 Analog Multiplexer/Demultiplexer

These are true multiplexers due to internal connections routing all external inputs to a common output. They function as digitally controlled analog switches featuring a low ON resistance and LOW internal leakage currents. *Figure 5.2.19* displays the package connections of the three types. The CD4051 is an 8-input, single output device. The CD4052 is two individual quad input switches. The CD4053 is a triple two-channel device. The figure

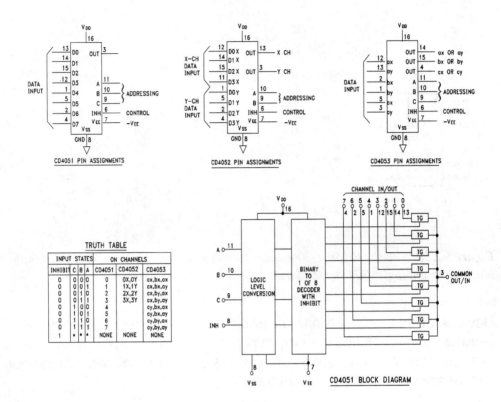

Figure 5.2.19. *The CD4051/CD4052/CD4053 device pin assignments, truth table, and CD4051 block diagram.*

includes the truth table for the three types. A block diagram for the C4051 is also provided.

Features common to the three device types include:

- Wide digital and analog signal ranges — digital of 3 to 15 volts, analog to 15 Vp-p. Note that the package provides connections for plus, (V_{DD}), and minus, (V_{EE}), power relative to ground (V_{SS}).
- Low ON resistance, 80Ω typical, over the entire permissible power and signal range.
- High OFF resistance, channel leakage of ±10 pA typical at $V_{DD} - V_{EE} =$ 10V.

- Logic level conversion for digital addressing of analog signals over the full range of permissible power and signal range. (Refer to the CD4051 block diagram in **Figure 5.2.19.**)
- Matched switch characteristics: $R_{ON} = 5\Omega$ (typ) $V_{DD} - V_{EE} = 15V$.
- Very low quiescent power dissipation under all permissible control and signal conditions: 1 µW (typ) $V_{DD} - V_{SS} = V_{DD} - V_{EE} = 10V$.
- On-chip binary address decoding.

Note that the CD4051 is used for input signal selection with the multichannel oscilloscope switch, **Figure 3.6.1.**

The Siliconix DG303A/DGP303A Dual Precision SPDT CMOS Analog Switch[3]

This device is unique in containing two sets of individually driven switches, one normally open, the other normally closed. By tying the two inputs of a switch pair together a SPDT configuration is achieved — or the two can be used individually with simultaneous state changes of the inputs. **Figure 5.2.20** provides a description of the packaging, a function diagram, and the truth table. Features unique to the DGP303A include:

- A ±22 volt input range.
- 10Ω rds(ON) for any combination of the switches.
- 0.5 nA max at 25°C, ±15V.
- Tested t_{OFF} and t_{ON} <50 nS.

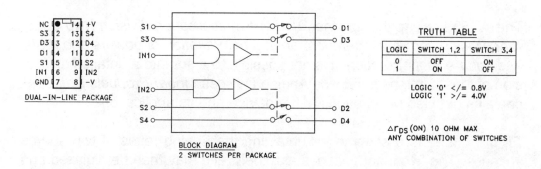

Figure 5.2.20. *The Siliconix DGP303A dual precision SPDT CMOS analog switch device pin assignments, block diagram, and truth table.*

3. *Siliconix Incorporated, Integrated Circuits Databook, 1988.*

An SPST switch, the DG300A, of similar specs to the DG303A is used with the *waveform burst* module.

5.2.6. The Real-Time Analog Multiplier/Divider

The AD538 Real-Time Computational Unit (ACU)

The AD538[1] is a computational circuit capable of precision analog multiplication, division, and exponentiation. The device features low input and output offset voltages and a high degree of linearity. Accurate computation is possible over a wide input dynamic range. Device features include:

- A dynamic range (denominator) of 1000:1.
- Simultaneous multiplication and division.
- Resistor programmable powers and roots.
- No external trims required.
- Low input offsets - <100 µV.
- Error ±0.25% of reading, 100:1 range.
- Two on-chip voltage references: +2 and +10 volts.

Figure 5.2.21 (a) shows the internal function diagram and DIP pin identification. The (b) portion of the figure is of the internal modeling process with equations. We see that the circuits operate by first converting the input voltages to their natural logarithmic values. This enables computation by simple op amp summing. The outputs then undergo an antilog process to restore their linearity.

The AD538 is capable of *one-quadrant multiplication/division, two quadrant division, log ratio operation, analog computation of powers and roots,* and *square root operation.* It is not possible to enter into a detailed discussion of these options here. The Analog Devices *Linear Products Databook* contains detailed descriptions of the device and its applications.

Figure 5.2.22 is a scheme for obtaining the analog ratios of two source voltages. The two input voltages and their ratio may then be digitized and displayed if desired.

---

1. *Analog Devices, Linear Products Databook, 1988.*

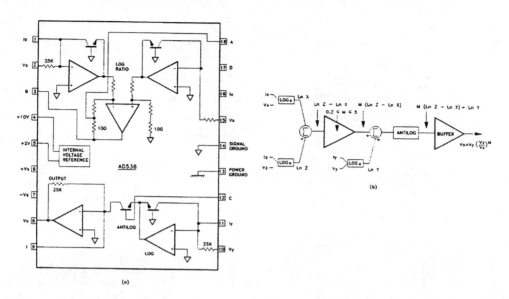

Figure 5.2.21. *(a) AD538 function diagram and 14-pin DIP numbering. (b) Internal circuit processing model.*

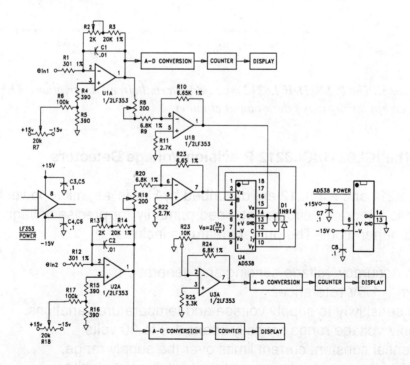

Figure 5.2.22. *A scheme for obtaining the ratio of two voltages using the AD538 in the two-quadrant division mode with 2-volt scaling.*

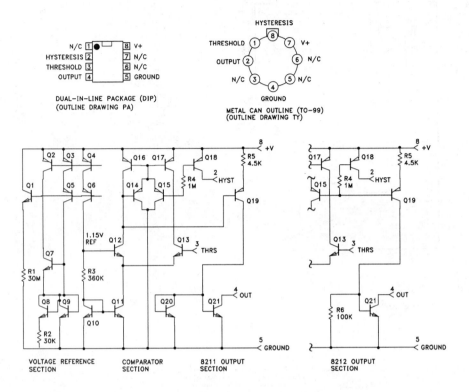

Figure 5.2.23. *The ICL8211/ICL8212 case and internal circuit descriptions. The two are identical except for details of their output circuitry.*

5.2.7. The ICL8211/ICL8212 Precision Voltage Detectors

The ICL8211 and ICL8212 are described by Harris[1] as "micropower bipolar monolithic integrated circuits intended primarily for precise voltage detection and generation." The primary features include:

• High accuracy voltage sensing and generation.
• Internal 1.5V reference.
• Low sensitivity to supply voltage and temperature variations.
• Supply voltage range typically spans 1.8 to 30 volts.
• Essential constant current limits over the supply range.
• Easily computed hysteresis values for set point stability.

1. Harris Semiconductor, LINEAR ICs For Commercial Applications, 1990, pp2-147 to 2-158.

I have made measurements of these which tend to confirm the data claims.

Figure 5.2.23 shows the case and internal circuit details. The case pin connections for the two devices are identical. We see from the circuitry that the *threshold* and *hysteresis* circuitry is identical for the two; the performance differences arise from the manner in which the open collector output is driven. This is shown individually for each device. It is these that account for the performance differences.

The key to the high performance is the 1.15 volt reference potential developed in the "voltage reference" section. This is made available at the *threshold* terminal, pin 3. The reference is extremely stable over the range of operating voltages and temperatures.

Looking at the 8211, the base of transistor Q21 is clamped by the drop across the diode connected Q20. The constant current diode drive from Q19 is derived from the 1.15 reference, making it extremely stable. The current sinking limit is about 7 mA: an LED may be driven directly from the power supply with no limiting resistor.

With the 8212, the diode connected transistor is replaced by resistor R6. Its high value of 100K forces most of the current from Q19 into the base of Q21. A much higher sink current is now available; an LED now requires a limiting resistor.

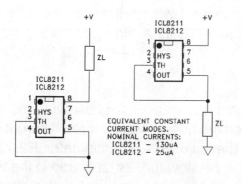

Figure 5.2.24. Constant current modes for the ICL8211 and ICL8212.

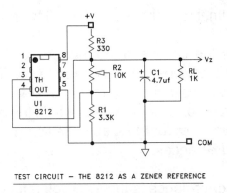

TEST CIRCUIT — THE 8212 AS A ZENER REFERENCE

MEASUREMENT DATA WITH PLUS V = 9, 12, AND 15 VOLTS

R2	9V	12V	15V
0	1.13	1.25	1.81
1000	1.51	1.52	1.82
2200	1.79	1.80	1.86
3300	2.15	2.15	2.17
4700	2.63	2.64	2.65
6800	3.43	3.44	3.45
10000	4.43	4.44	4.45

NOTE: THERE WAS NO SIGNIFICANT CHANGE IN VZ WITH RL DISCONNECTED

Figure 5.2.25. The measurement circuit I employed to assess the ICL8212 as a high performance zener reference.

Figure 5.2.24 illustrates the two ICs in a constant current mode. The current load may be in either of the two locations shown. The constant current values in the figure are from the Harris data book. My own measurements show fairly close correspondence with several 8212s; somewhat less with the 8211s; which I found surprising.

I am not too enthused over the application of these as a constant current source. For most purposes I would prefer to use a JFET or a transistor collector. However, the 8212 as a high performance zener reference is, from my measurements, outstanding — thus its use in *a precision constant voltage/current source* (Chapter 3.3).

Figure 5.2.25 diagrams a test circuit for observing the zener performance of the 8212. I observed the output, Vz, for three different supply voltages: 9, 12, and 15 volts. For these I used a potentiometer, R2, ranging from zero to 10,000 ohms to set the zener value, as tabulated in the figure. With R2 less than 2200 ohms the variation was significant; with higher loading the stability was excellent over the range of voltages. The 1000 ohm load was not material: disconnecting the resistor had no effect on the zener output.

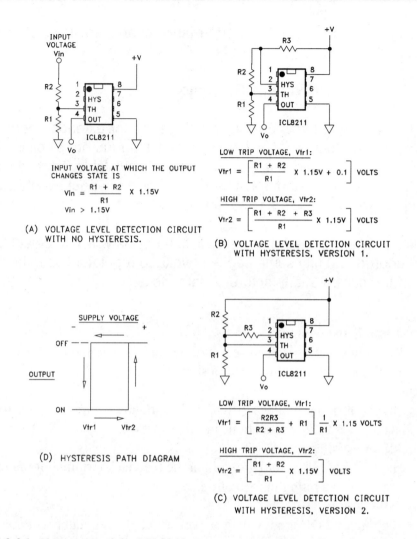

Figure 5.2.26. Variations of the ICL8211 as a programmable voltage level detector. The (C) version is recommended for low voltage applications.

Figure 5.2.26 describes the 8211 in its role of low voltage detector. As seen in the figure, the input source may be an independent voltage or the supply itself. Hysteresis may, but need not, be provided. In general the use of hysteresis is preferable to maintain stability of the transition points.

There are many possible applications for the 8211/8212. Their utility makes it worthwhile to peruse the Harris data. Before deciding on an application it is wise to review the abundant performance characteristics graphically described.

5.2.8. The Linear IC Voltage Regulator

Back in the 1960s, I was developing a system for extended operation on the deep ocean floor. Alkaline batteries with a startup of 36 volts, which fell off rapidly with use, were the best power source available. To get the needed operating time, the system had to function down to about twenty volts. Designing the regulator was an interesting challenge.

The times have changed and the question now is which to select from the three hundred and fifty some pages of voltage regulator ICs to be found in some of the current manufacturers' databooks.

Current-Limit Sense Voltage:
> The voltage across the current limit terminals required to cause the regulator to current-limit with a short circuited output.

Dropout Voltage:
> The input-output voltage differential at which the circuit ceases to regulate against further reductions in input voltage.

Feedback Sense Voltage:
> The voltage, referred to ground, on the feedback terminal of the regulator while it is operating in regulation.

Input Voltage Range:
> The range of DC input voltages over which the regulator will operate within specifications.

Line Regulation:
> The change in output voltage for a change in the input voltage.

Load Regulation:
> The change in output voltage for a change in load current at constant chip temperature.

Long Term Stability:
> Output voltage stability under accelerated life-test conditions at 125°C with maximum rated voltages and power dissipation for 1000 hours.

Maximum Power Dissipation:
>   The maximum total device dissipation for which the regulator will oper-
>   ate within specifications.

Output-Input Voltage Differential:
>   The voltage difference between the unregulated input voltage and the
>   regulated ouput voltage for which the regulator will operate within
>   specifications.

Output Noise Voltage:
>   The RMS ac voltage at the output with constant load and no input ripple,
>   measured over a specified frequency range.

Output Voltage Range:
>   The range of regulated output voltages over which the specifications
>   apply.

Quiescent Current:
>   That part of the input current to the regulator that is not delivered to the
>   load.

Ripple Rejection:
>   The line regulation for ac input signals at or above a given frequency
>   with a specified value of bypass capacitor on the reference bypass
>   terminal.

Standby Current Drain:
>   That part of the operating current of the regulator which does not
>   contribute to the load current.

Temperature Stability:
>   The percentage change in output voltage for a thermal variation from
>   room temperature either temperature extreme.

Thermal Regulation:
>   Percentage change in output voltage for a given change in power
>   dissipation over a specified time period.

Table 5.2.6. Voltage regulator definition of terms.

Actually, I have found that as a hobbyist there are a few basic types that meet the majority of my needs; but every so often something out of the ordinary sends me back to the databooks. To make life easier this section provides an overview of most of the currently available regulators. By overview is meant a selection guide listing attributes and a drawing illustrating

(a) Adjustable Positive Voltage Regulators

AMPS	DEVICE	OUTPUT VOLTS	PACKAGE
10.0	LM396K	1.25-15V	TO-3
5.0	LM338K,T	1.2-32V	TO-3,TO-220
3.0	LM350K,T,AT	1.2-33V	TO-3,TO-220
1.5	LM317K,T,AK,AT	1.2-37V	TO-3,TO-220
1.5	LM317HVK	1.2-57V	TO-3
0.5	LM317H	1.2-57V	TO-39
0.5	LM317HVH	1.2-37V	TO-39
0.5	LM317MP	1.2-37V	TO-202
0.1	LM317LZ,M	1.2-37V	TO-92,SO-8
0.1	LM2931CT	3.0-24V	TO-220,5-LEAD

(b) Adjustable Negative Voltage Regulators

AMPS	DEVICE	OUTPUT VOLTS	PACKAGE
3.0	LM333K,T	-1.2-32V	TO-3,TO-220
1.5	LM337K,T	-1.2-37V	TO-3,TO-220
1.5	LM337HVK	-1.2-47V	TO-3
0.5	LM337H	-1.2-37V	TO-39
0.5	LM337HVH	-1.2-47V	TO-39
0.5	LM317MP	-1.2-37V	TO-202
0.1	LM317LZ,M	-1.2-37V	TO-92,SO-8

(c) Adjustable Positive/Negative Voltage Regulators

AMPS	DEVICE	OUTPUT VOLTS	PACKAGE
±0.1	LH7001	±1.2-37V	DIP,TO-5

(d) Fixed Positive Voltage Regulators

AMPS	DEVICE	OUTPUT VOLTS	PACKAGE
3.0	LM323K,AK	5V	TO-3
1.0	LM309K	5V	TO-3
1.0	LM340AK,T	5,12,15V	TO-3,TO-220
1.0	LM340K,T	5,12,15V	TO-3,TO-220
1.0	LM78xxCK,T	5,6,8,12,15,18,24V	TO-3,TO-220
0.5	LM2984CT	5,12,15V	TO-220,TO-202
0.5	LM341T,P	5,12,15V	TO-220,TO-202
1.0	LM78xxCK,H	5,6,8,12,15,18,24V	TO-3,TO-39
0.2	LM309H	5V	TO-39
0.2	LM342P	5,12,15V	TO-202
0.15	LM2930T	5,8V	TO-220
0.15	LM330T	5	TO-220
0.1	LM340LZ,H	5,12,15V	TO-92,TO-39

AMPS	DEVICE	OUTPUT VOLTS	PACKAGE
0.1	LM78xxACZ,H,M	5,6,8,12,15,18,24V	TO-3,TO-39, SO-8
0.05	LM2936Z	5V	TO-92

(e) Fixed Negative Voltage Regulators

AMPS	DEVICE	OUTPUT VOLTS	PACKAGE
3.0	LM345K	-5,5.2V	TO-3
1.5	LM320K,T	-5,12,15V	TO-3,TO-220
1.5	LM79xxCT,K	5,6,8,12,15V	TO-3,TO-220
0.5	LM320MP	-5,12,15V	TO-220
0.5	LM79xxCT,H	-5,8,12,15V	TO-39,TO-220
0.2	LM320H	-5,12,15V	TO-39
0.1	LM320LZ	-5,12,15V	TO-92
0.1	LM79xxACZ,M	-5,12,15V	TO-92,SO-8

(f) Low Dropout Regulators

AMPS	DEVICE	OUTPUT VOLTS	PACKAGE
0.050	LM2936Z	5V	TO-92
0.100	LM2931T	5V,ADJ	TO-220,TO-92
0.100	LM2950CZ	5V	TO-92
0.100	LM2951N,J,H	ADJ	DIP, CERDIP, HDR
0.150	LM2930T	5,8V	TO-220
0.750	LM2935T	DUAL 5V	TO-220, 5-LEAD
1.00	LM2940T	5,8,10,12V	TO-220
1.00	LM2940CT	5,12,15V	TO-220
1.00	LM2941T	5-20V,ADJ	TO-220
1.00	LM2941CT	5-20V,ADJ	TO-220

(g) Shunt Regulators

AMPS	DEVICE	OUTPUT VOLTS	PACKAGE
0.15	LM431ACZ,M	2.5-36V	TO-92,SO-8

(h) Building Block Regulators

DEVICE	TITLE	PACKAGE
LM104/204/304	Negative Regulator	TO-39
LM105/205/305	Positive Regulator	TO-39
LM376	Positive Regulator	8-Pin Plastic DIP
LM723	Voltage Regulator	14-Pin DIP,TO-39

Table 5.2.7. Voltage regulator selection guide, commercial temperature range devices.

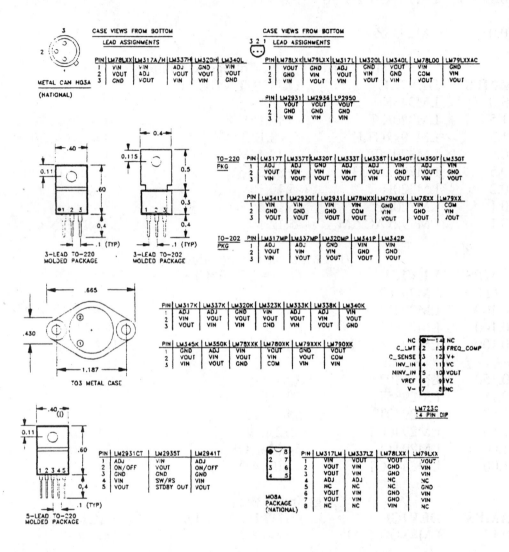

Figure 5.2.27. A summary of the available voltage regulator device packages.

the range of pacakge styles with a summary of regulators availabel for each.
Keep in mind these are strictly guides — there is no way to list all the attributes of the vast number available. Use the guides to narrow the search,
then reach for a databook.

Advantages of currently available types include:

- Compact and extremely rugged packaging with excellent thermal conduction properties.
- Internal overload protection, including output shorting.
- Thermal overload shutdown.
- High stability with a minimum of filtering.
- Virtually rock solid regulation over a large range of source and input loading.
- Low dropout limits, some down to 0.6 volts.

The availability of compact packaging with excellent thermal characteristics has made IC placement on the PC board it powers practical with attendant savings in space and cost.

Table 5.2.6 provides a definition of regulator terms. It is worthwhile to become acquainted with these, in particular those having to do with source and load regulation, dissipation, and stability.

Table 5.2.7 is a selection guide of commercial temperature grade regulators by function and current ratings. *Figure 5.2.27* describes the various packages with a listing of related types.

5.3. DIGITAL LOGIC

5.3.1. Logic Families

The TTL Logic Family

The transistor-transistor logic (TTL) family was not the first of the digital integrated circuits to hit the market, but it is the logic family that set the standard. TTL logic was preceded by several other types: *direct-coupled transistor logic* (DCTL), *resistor-transistor logic* (RTL), *resistor-capacitor-transistor logic* (RCTL), and *diode-transistor logic* (DTL). Of these, as I recall, DTL was the most popular. Another bipolar logic family that followed TTL is *emitter coupled logic* (ECL)[1]. This family is known for its very fast

---

1. *For a discussion of these an excellent review is provided in A User's Handbook of Integrated Circuits, Eugene R. Hantek, John Wiley & Sons, 1 973.*

switching speed, just a few nanoseconds, along with a relatively high power dissipation. Unfortunately, there is a relationship between these — in general, the price of higher speed is greater dissipation.

While proprietary TTL circuitry was developed by various manufacturers — Fairchild's 9300 and Signetics 8200 series come to mind — the 5400 series, pioneered by Texas Instruments Incorporated for military use, quickly assumed the status of a de facto standard. The commercial version, designated 7400, soon followed[2, 3]. The TTL family has also spawned offspring with the twin objectives of higher speed and lower power requirements, not necessarily in the same devices. These include:

- A higher power, higher performance family, the 54H/74HOO series. Basically the same as the standard series but having lowered resistor values and clamping diodes.
- Schottky clamped transistor 54S/74SOO, high power, high speed devices.
- A low power series, 54L/74LOO, designed for improved speed and power, predating the 54LS/74LSOO series.
- Low power Schottky, 54LS/74LSOO, featuring about the same performance of the 54/7400 type but with about 1/4 the power requirement.
- Advanced Schottky/advanced low power Schottky, 54AS/74ASOO, 54ALS/74ALSOO pin and functionally compatible with existing 54/74xx series of deviceS[4].
- 74FOO, introduced by Signetics as "the high speed logic of the 1980s."[5].

The following are common to all members of the family:

- Supply voltage 4.75 to 5.25 volts.
- Logical 0 (LOW) output voltage 0.4 volt.
- Logical 1 (HIGH) output voltage 2.4 volt.
- Noise immunity 0.4 volt.

2. IC Applications Staff, Robert L. Morris and John R. Miller, Editors, Designing With TTL Integrated Circuits, Texas Instruments Incorporated, McGraw-Hill Book Company, 1 97 1.
3. A fascinating book providing early IC design background is Application Memos, Signetics Corporation, September 1 969.
4. Texas Instruments, ALS/AS Logic Data Book, 1986
5. Signetics Corporation, Fast Data Manual, 1987

- Commercial grade operating temperature range O'C to 70'C.
- Military grade operating temperature range -55'C to +125'C.

Figure 5.3.1 contrasts the basic DTL circuit with the TTL. These two logic types operated in a similar manner, but the unique multiple emitter of the input transistor and enhanced output structures stole the show for TTL.

The *totem pole* output shown in the drawing is the original standard. An alternative is to omit the pullup transistor and diode, leaving the lower open collector (OC) transistor as the output. This is advantageous for connecting several outputs in parallel, referred to as the wired OR connection. The open collector power is independent of Vcc and for some devices may exceed five volts. Some examples are:

OC Device	VMax	OC Device	VMax	OC Device	VMax
7403/09/15/38	5V	7416/26	15V	7406/07	30V

Analog signals are far more sensitive to circuit noise and voltage drifts than digital. Digital voltages are defined by logic levels — 0 meaning LOW and 1 meaning HIGH. The actual voltage corresponding to these depends on the logic device. For TTL logic LOW is a voltage ranging from 0 to +0.4 volts; logical HIGH ranges from +2.4V to Vcc. Additionally, a noise margin of 0.4 volts is specified; that is, a LOW may be as high as 0.8V and a HIGH as low as 2.0V and still be within tolerance.

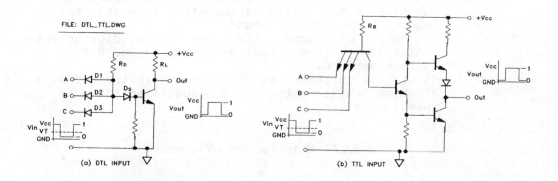

Figure 5.3.1. DTL and TTL logic circuit 3-input gate comparison.

```
Term/Symbol  Definition
-----------------------------------------------------------------
  Vcc        Supply Voltage. The range over which the device is
             guaranteed to operate within specified limits.
             Initially applied to TTL, currently relates to CMOS
             as well as an alternative to VDD.
  VDD        Supply Voltage. Originally applied to MOS devices,
             then CMOS positive power.
  Vss        MOS negative, also CMOS ground.
  VIH        Input HIGH voltage. The range of input voltages
             recognized by the device as HIGH.
  VIH(Min)   The minimum allowed input HIGH.
  VIL        Input LOW voltage. The range of input voltages
             recognized by the device as LOW.
  VIL(Max)   The maximum allowed input LOW.
  VOH(Min)   Output HIGH voltage. The minimum guaranteed HIGH
             voltage at an output terminal for the specified
             output current and the minimum Vcc value.
  VOL(Max)   Output LOW voltage. The maximum guaranteed LOW
             voltage at an output terminal sinking the specified
             load current.
  Icc        Supply Current. The current flowing into the Vcc
             terminal of the device.
  Ii         Input leakage current. The current flowing into the
             input at the maximum voltage.
  IIH        Input HIGH current. The current flowing into the
             input at the specified HIGH input voltage.
  IIL        Input LOW current. The current flowing out of a
             terminal when a specified LOW voltage is present.
  IOH        Output HIGH current. The leakage current flowing into
             a turned off collector with specified HIGH voltage
             present.
  IOL        Output LOW current. The current flowing into an
             output which is at the LOW state.
  Ios        Output short-circuited current. The current flowing
             out of an output which is the HIGH state when the
             output is shorted to ground.
  IOZH       The output off current HIGH. The current flowing into
             a disabled 3-state output with a specified HIGH
             voltage applied.
  IOZL       Cutoff off current LOW. The current flowing out of a
             disabled 3-state output with a specified LOW output
             voltage applied.
```

Table 5.3.1. DC symbols and definitions.

Output loading constraints must be observed. In the HIGH state the circuit sources current to the load. In the LOW state the output is a current sink.

Term	Definition
f_{MAX}	Operating frequency. The fastest rate at which a circuit may be toggled.
t_{PHL}	Propagation delay from input to output, output going, high to low.
t_{PLH}	Propagation delay from input to output, output going, low to high.
t_{pZH}	Enable propagation delay time. This is measured from the input to the output going to an active high level from three-state.
t_{pZL}	Enable propagation delay time measured from the input to the output going to an active level from three-state.
t_{pHZ}	Disable propagation delay time to the output going from an active high level to three-state.
t_{pLZ}	Disable propagation delay time to the output going from an active lowlevel to three-state.
t_W	Input signal pulse width.
t_S	Input setup time, the time that data must be present prior to clocking input transitioning.
t_H	Input hold time, the time that data must after clocking input has transitioned.
t_{REM}	Clock removal time, the time that an active clear or enable signal must be removed before the clock input transitions. Sometimes call the recovery time.
t_r	Input signal rise time.
t_f	Input signal fall time.
t_{TLH}	Output rise time, transistion time from low to high.
t_{THL}	Output fall time, transistion time from high to low.

Table 5.3.2. AC parameter definitions.

Fanout is a term defining the number of logic loads that may be driven from a given output. A standard TTL output will source a minimum of 400 µA and sink a minimum of 16 mA. An input will source a maximum of 1.6 mA when pulled low and sink a maximum of 40 µA when pulled up. This equates to a fanout of ten standard loads. However other factors exist, primarily device temperature and transition times, which are affected by capacitive loading. 74LS logic currents are one-half that of standard 7400 TTL. Thus a 74LSOO gate can drive but five standard 7400 inputs.

Table 5.3.1 defines device DC terminology; *Table 5.3.2* defines AC terminology. For the most part these apply equally to the TTL and CMOS series. Where a distinction exists it is noted.

The CMOS Logic Family

The circuits I design in my home lab and have entered in this book use CMOS exclusively. This is because CMOS comes very close to being an ideal logic family[6].

An ideal logic family would dissipate no power and experience no internal propagation delays. Static dissipation is on the order of l0 nW per gate, largely due to leakage currents. Active power dissipation is dependent on several factors: the supply voltage, signal frequency, output loading, and input rise time. Typical gate dissipation at I MHz with a 50 pF load is said to be less than 10 mW.

CMOS rise and fall times appear to be ramps and 20 to 40 times longer than the internal propation delays: on the order of 25 to 50 nS for a typical gate. The ramp results from the extremely high input resistance shunted by a few pF of capacitance. Noise immunity approaches fifty percent, typically being 45% of the full logic swing.

When CMOS devices first came on the scene back in the 170s they were costly compared to TTL and very sensitive to improper handling. Considerable improvement has been made in both respects. While careful handling is still required, internal protection has greatly reduced the risk[7].

Figure 5.3.2 shows the circuitry for a two-input OR gate. The figure also illustrates diode arrangements that afford input protection. The basic CMOS building block is the inverter. This circuit contains five: one for each input and three following to the output. Note the simplicity inherent in the design — repeated structures with no intervening resistors or other components.

The inverter consists of two parallel connected enhancement mode MOSFETs. The upper, QI, is a P-channel type; the lower, Q2, an N-channel. When the gate-to-source voltage is zero ($V_{GS} = O$), the device is cut off

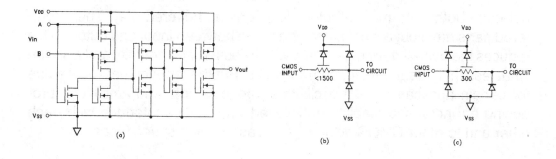

Figure 5.3.2. (a) 2-input CMOS OR-logic gate circuit. (b) Double diode input protective circuit. (c) Triple diode input protective circuit.

— the drain-to-source conduction is zero. we note that the inverter's input and output connections are at the midpoint of the two MOSFETs; which is to say, at Vcc/2 voltage-wise. If the input, V_{IN}, rises, and Ql is cut off while Q2 is driven to saturation, forcing the output to ground. Conversely when the input goes low, Q2 is cut off and the output is pulled up to Vcc. Well, almost. V_{OUT} ideally swings fully between ground and the positive supply, but reality stops it just a bit short — but not by much. The inverter input resistance is on the order of a trillion ohms, shunted with a small capacitance, 5 pF typically. So the logic levels are essentially the limits of the power supply.

In the beginning, CMOS power supplies were called VDD and Vss, carried over from MOS circuits where they refer to Drain and Source, respectively; but the TTL Vcc and ground convention persisted and we see both in use now. My own preference is for VCC and Ground. The original CMOS devices operate with a wide range of DC power but the later high speed series developed for TTL interfacing have a more limited range.

When RCA introduced the CMOS line the series was given the number 4000 preceded by CD. A number of manufacturers now provide the devices and as with TTL the CMOS family tree has developed a series of advanced variations on its branches. A leap in numbering was made from 40xx to 45xx as the range of functions expanded. Motorola inserted a "MCI" prefix, i.e., MC14x00. Also, devices are defined as *buffered* (B), or *unbuffered* (UB). A glance at a selection guide shows most gates to be designated as both, with most all other functions as buffered only. The unbuffered parts have but a single inverting stage between input and output. This reduces the overall device gain, and with it, less noise immunity and a somewhat less than ideal transfer characteristic. But the reduced gain is of value for certain applications like oscillators, monostables, and amplifiers, producing an increased stability and speed. Both types interface with each other and to other CMOS devices, such as the high speed types.

In the early 1970s National Semiconductor introduced a series 54/74 CMOS line as a pin-for-pin replacement for 54/74 TTL. Well, almost all; I got caught by an overlooked switch of two of the four gates of the 74C86 vs. the 7486. The two CMOS families can be mixed, in fact the National CD4069 and 74CO4 hex inverter carries both numbers. National has an application note number 90, *54CI74C Family Characteristics*[6], which defines ail the characteristics common to both families. The National CMOS Logic Databook contains 29 excellent informative application notes. There are several manufacturers of CMOS ICs. Though they are similar, small differences frequently exist. The only safe way to work with these is to check out the maker's specs.

Two follow-on series to the 54C/74C line are the high speed 54HC/74HC and its sub-family 54HCT/74HCT series[7]. While these are logically compatible there are significant differences we need to be aware of. The HC series carry the same pinouts and functions as the 54LS/74LS series. The 54HCT/74HCT duplicate the 54LS/74LS by one step further in operating at TTL logic levels. If we wish to use an HC, HCT, or AC/ACT device we must consider the device's Vcc specifications.

DC power ranges for various series (all voltages positive with respect to ground) are:

CD400OB/UB	5-15V	CD400OM/C	3-15V
74COO	3-15V	74HV	2-6V
74HCT	4.5-5.5V	74AC/ACT	0.5-6V
MM74C912	-MM74956	3-6V	

Crossvolt[8] Low Voltage Logic Series

At this writing there are six members of this family from National Semiconductor Corporation of which I am aware. The six are:

- LVQ - Low voltage quiet CMOS logic.
- LVX - Low voltage quiet CMOS logic with 5-volt tolerant inputs.
- LVX - Low voltage dual supply CMOS translating tranceivers.
- LVX - Low voltage CMOS bus switches.
- LCX - Low voltage high speed CMOS logic with 5-volt tolerant inputs and outputs.
- LVT - Low voltage high speed BICMOS logic.

The devices all share the following features:

- Low or "zero', static power dissipation (<l00 nA typical for LVQ).
- Reduced dynamic power consumption.
- Lower switching noise than comparable higher supply voltage counterparts.
- Compliance with EIA-JEDEC low voltage interface standard #B-lB.

Each member of the series family possess a unique set of features and operating characteristics optimized for a specific low voltage application. Output drive, switching speed, translation capabilities, and interface flexibility are differentiating characteristics of these families.

5.3.2. Basics of Digital Logic

Number Systems

Hey! This is a book on electronics. Number systems are for computers. Let's stick with the subject.

Well, we are. The fact is, the foundation for digital electronics is based on the binary number system — the system we would quite likely be using if we had but two fingers and toes. So it is to our advantage to make its acquaintance.

Binary	Octal	Decimal	Hexadecimal
0000	0	0	0
0001	1	1	1
0010	2	2	2
0011	3	3	3
0100	4	4	4
0101	5	5	5
0110	6	6	6
0111	7	7	7
1000	10	8	8
1001	11	9	9
1010	12	10	A
1011	13	11	B
1100	14	12	C
1101	15	13	D
1110	16	14	E
1111	17	15	F

Table 5.3.3. The relationship between binary, octal, decimal, and hexadecimal numbers.

In our everyday world we mostly use a number system based on the digits 0 through 9; but just try to imagine maintaining ten stable, separable voltage levels over an indefinite length of time with drifting power supplies and all the environmental factors. I understand it has been tried.

In the present-day digital circuits of interest only two voltage levels are required. Designated LOW and HIGH, these are *logic* values which relate to *physical* values (voltages) by the two numbers 0 and 1. This scheme allows us to discuss logic systems regardless of the voltages in use. I must admit that in my introduction to digital I had trouble comprehending the concept; but then, in those early days, there were about as many voltages, both positive and negative, as there were system designers.

Somehow these two numbers must be made to carry all the computing and control functions needed in a digital system. This is typically done by assigning the value 0 to circuit ground and 1 to a voltage other than ground. For most of our digital activity this will be a low valued positive voltage. Logic designations for 0 are LOW and FALSE. Designations commonly

Decimal $9635 = 9*10^3 + 6*10^2 + 3*10^1 + 5*10^0$

Now let's convert 9635 to its binary equivalent. To do this we perform successive divisions by 2 and keep track of the remainders, which will constitute the binary value.

```
9635/2 = 4817 + 1    LSB
4817/2 = 2408 + 1
2408/2 = 1204 + 0
1204/2 =  602 + 0                   Conversion to Hexadecimal:

 602/2 =  301 + 0                   10 0101 1010 0011 BINARY
 301/2 =  150 + 1                    2   5    A    3  HEXADECIMAL
 150/2 =   75 + 0
  75/2 =   37 + 1

  37/2 =   18 + 1                   Conversion to Octal:
  18/2 =    9 + 0
   9/2 =    4 + 1                   10 010 110 100 011 BINARY
   4/2 =    2 + 0                    2   2   6   4   3 OCTAL

   2/2 =    1 + 0
   1/2 =    0 + 1    MSB
```

Table 5.3.4. Conversion of a decimal number to its binary, hexadecimal and octal equivalents.

used for 1 are HIGH and TRUE. By thinking of the logic terms we can avoid ties to any specific voltage.

In our minds we see the binary values 0 and 1 as two possible states of existence. We can then utilize the concept in a class of devices that may have two or more inputs operating to control but a single output. Each input and the output can be in one or the other of the two possible states. The state of the output will always be dependent on some combination of the input states. For this reason we use the expression combinational logic when describing the action of logic gates.

We're getting ahead of ourselves, so let's focus now on what we mean by number systems. Number systems are defined by a base number, or *radix*.

```
        Binary Addition                    With Extended Carries
        ---------------                    ---------------------------
Augend + Addend = Sum + Carry      Dec  Binary   Dec    Binary
  0    +    0   =  0  +   0          4     100    109    1101101
  0    +    1   =  1  +   0         +5     101    +91   +1011011
  1    +    0   =  1  +   0         --    ----    ---   --------
  1    +    1   =  0  +   1          9    1001    200   11001000


  One's Complement                   Two's Complement
  ----------------                   ----------------
   11001100101001                     00110011010110
   00110011010110                                 +1
                                     ---------------
                                      00110011010111


  Subtraction                        Subtraction with 2's complement
  -----------                        -------------------------------
    100011                            100011  Minuend
   -010101                           +101011  2s Complement of Subtrahend
  ---------                          -------
    001110                            001110
                         discard>1
```

Table 5.3.5. Binary addition, subtraction with ones and twos complement examples.

The base for our decimal system is the number ten. The base for binary numbers is two. We see that any number could be used as a base. There are only four that are of interest to us however. These are the *binary*, base two; the *octal*, base 8; *decimal*, base 10, and *hexadecimal*, base 16. **Table 5.3.3** illustrates the relationships between these. The octal series may seem confusing with its "10" corresponding to the decimal "8." The term carry relates to the digit "1" that is carried to the left at each multiple of the base number. Thus the next number in the *hexadecimal* column is "IF" followed by "2F," "3F," and so on at each multiple of 16. The value 256 in decimal is FF in hexadecimal, which is 16 times 16 in both systems.

Suppose we look at how the four number systems relate to each other. We see this in **Table 5.3.4**. I deliberately use a large decimal number as I feel this brings about a better understanding than the shorter numbers found in many texts.

When we look at the four-digit number, 9635, we see it as the sum of each of its four components taken to an ascending power 10. Binary numbers have a similar makeup although its components are taken to the power of 2. Similarly, octal numbers are summed as multiples of powers of 8 and hexadecimal as powers of 16.

The conversion of a decimal number to binary consists of successive divisions by 2 with the quotient and remainder separated as shown. All even numbers have a remainder of 0, all odd a remainder of 1. Note the assignment of LSB at the head of the column and MSB at the base. These stand for *least significant bit* and *most significant bit*, respectively. As one speaking from experience it is all too easy to switch these around.

So why use *bit* rather than *digit*?

In the language of the computer world individual zeros and ones are called *bits*. Eight bits taken together are a *byte*. (I have seen four bits called a *nibble*.) Electrically these bits and bytes are digital signals, which we enter into gates and flip-flops and get back from counters and registers. We use divide-by-10 counters for decimal functions and divide-by-16 for binary — some, such as three-state buffers, as octal.

Table 5.3.5 delves into the mysteries of binary arithmetic. Binary arithmetic is no different from decimal in principle, just more confusing. There are functions we can carry out in binary that we are unlikely to attempt with decimal: complements. Complementing a binary bit is analogous to inverting the polarity of an electrical bit.

There are two kinds of complement: ones and twos. Their operation is shown in the table. The complements play a key role in bipolar A/D and D/A conversions.

5.3.3. Combinational (GATE) Logic

In everyday thinking a gate is an entrance or passage. A logic gate is a passage for binary information. When the combination of low and/or high states at the inputs is correct the gate carries that information through to its output. In general there is more than one combination of gate inputs that result in a TRUE output.

Switching Algebra[1]

The mathematical definitions for logic gates are based on Boolean Algebra, often simply called switching algebra. The three basic operations are the AND (•), the OR (+), and the NOT (¯) functions. The values in parenthesis are symbols used in equations describing the operations. The "-" and "+" operators as used here do not represent the familiar arithmetic functions. Negation of a function is shown by a line drawn above the term(s). The three basic operations can be operated on in turn to create three additional functions: the *NAND*, the *NOR*, and the *exclusive OR* (EX-OR). **Figure 5.3.3** illustrates these six basic gates. Though only two gate inputs are shown, the rules apply for any number. The small circle on the output indicates it to be the negative of its positive counterpart. Note that an input can be negated as well as an output. Thus the inverter could be drawn with the circle at either port without affecting its meaning. The figure includes a new feature, the *truth table*. The truth table is a means of describing all the possible relationships between the input and output of the gate. Though the figure only shows two input gates the logic holds for any number of 4 inputs. We can show this with the truth table for the AND and NAND functions of a three-input gate:

Inputs A B C	AND Output F	NAND Output F
0 0 0	0	1
0 0 1	0	1
0 1 0	0	1
0 1 1	0	1
1 0 0	0	1
1 0 1	0	1
1 1 0	0	1
1 1 1	1	0

1. *A good reference text for all of section 5.3 is The Design of Digital Systems by John B. Peatman, McGraw-Hill Book Company, 1972.*

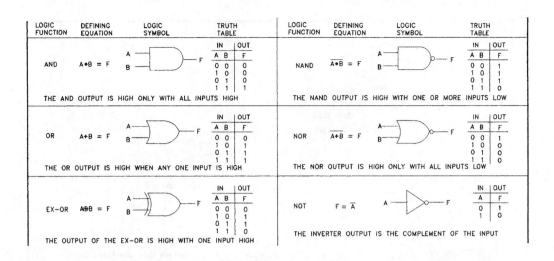

Figure 5.3.3. *Logic definitions, defining equations, and symbols for the six basic logic gates.*

The equations for the outputs are (AND) F = (A • B • C) and (NAND) F = ($\overline{A}$ • B • C).

Boolean Functions

To be useful, differing gate types must commonly be combined to form a required output. This often requires some imaginative thinking as well as mathematical juggling acts. There are some basic rules that are of help in many situations.

Figure 5.3.4 illustrates AND and OR logic concepts by a switching analogy. The expansion to more complex logic is shown by developing a functioning exclusive-OR gate with inverter and NAND construction.

Reference is made in the figure to *DeMorgan's theorem*. The theorem is a valuable aid in obtaining a required output from a set of known inputs. The basic form is:

$$\overline{A \bullet B} = \overline{A} + \overline{B}, \ \overline{A + B} = \overline{A} \bullet \overline{B}$$

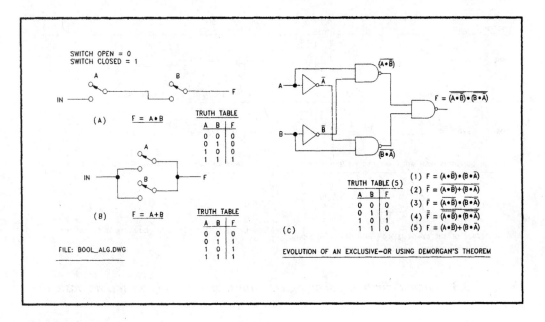

Figure 5.3.4. Switching logic basics: (a) AND logic. (b) OR logic. (c) Derivation of a 2-input exclusive-OR gate from inverter and NAND components.

This tells us that when we invert a logic function linked by an AND or OR operator, we invert its component functions and the operator. We saw an example transformation in the figure.

Just as with ordinary algebra, there are basic rules for Boolean. Some of these are:

1. $A + O = A$
2. $A + I = 1$
3. $A + A = A$
4. $A \cdot 0 = 0$
5. $A \cdot 1 = A$
6. $A \cdot A = A$

The commutative, associative, and distributive laws also apply:

7. $A + B = B + A$
8. $A \cdot B = B \cdot A$

9. ABC = AB(C) = B(AC) = C(AB)
10. AB + AC = A(B + C)
11. A + $\overline{A}$ = 1
12. A • $\overline{A}$ = 0

As an example, show that:

13. A(A+B) = A   AA + AB = (A + AB) = A(I + B) = A

It is true these relationships had more application in earlier times when medium and large scale ICs were not yet available; but it is to our advantage to know of their existence and to make use of them when the situation warrants.

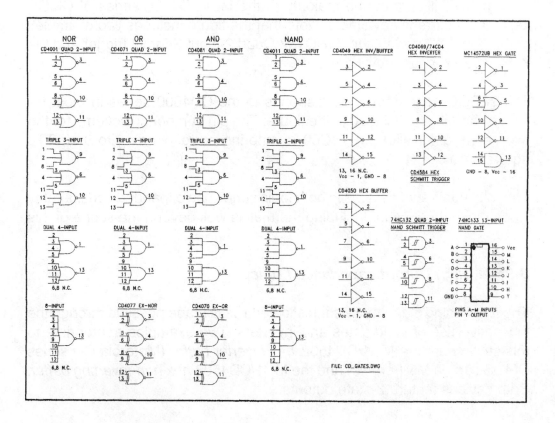

Figure 5.3.5. CD4000 series of CMOS gates and inverters.

Switching Function Simplification

Two methods of simplification are commonly used: a minimum *sum of prod-ucts* (disjunctive) or *product of sums* (conjunctive). An example of a sum of products form is:

$$F = \overline{A}\overline{B}\overline{C} + \overline{A}BC + \overline{A}B\overline{C} + A\overline{B}\overline{C} + AB\overline{C}$$

An example of a product of sums form is:

$$F = (A + \overline{B} + C)(\overline{A} + \overline{B} + C)(\overline{A} + \overline{B} + \overline{C})$$

Some Available CMOS Logic Gates

Figure 5.3.5 illustrates the majority of the MCI/CD4000 series of CMOS logic gates currently available. Note that the figure includes two examples of the 74HCOO series; the 74HC132 Quad 2-input Schmitt trigger and the 74HC133 13-input NAND gate.

The 74COO series gates corresponds to the CD4000 series in most re-spects though not included in the figure. The pin numbering scheme for the two series differ with the 74C00 numbering corresponding to their TTL equivalents.

The total number of devices in both logic series has expanded considerably since their introduction. Including them all is well beyond the scope of this book.

3-State, High Output Impedance Logic

The functioning of the more sophisticated ICs is made possible through the innovative use of logic gates and inverters. Two examples appropriate to this section are the MC1/CD4502 *hex inverter buffer* (Motorola CD series CMOS carry a MC1 prefix.) and the MC1/CD4503 *hex non-inverting buffer*. Both devices feature 3-state outputs.

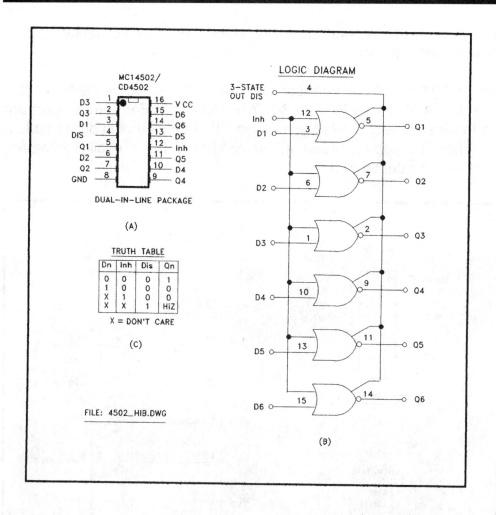

Figure 5.3.6. *The CD4502/MC14502 hex inverter 3-state buffer. (a) The dual-in-line package. (b) The device internal logic diagram. (c) The device truth table.*

The 3-state output is just that: presenting LOW, HIGH and HIGH IMPEDANCE (HiZ) logic levels to the outside world. 3-state devices were initially marketed by *National Semiconductor Corporation* under their trademark designation TRI-STATE. This feature places the output resistance beyond that of the normal HIGH impedance such as to approach an infinite value. The advantage is that any number of outputs can be parallel connected. It is, of course, essential that one, and only one, output be active at any given time. An enabling/disabling terminal is provided for this purpose. This terminal should not be confused with the inhibit function which relates to control of the internal logic.

Figure 5.3.6 shows the logic configuration of the MCl/CD4502/MC14502. Note the terminal identified as 3-STATE OUT DIS. For normal operation this terminal is held LOW. When raised HIGH the output is pulled up to a much higher impedance level. In this condition it has a negligible loading effect on the external circuitry.

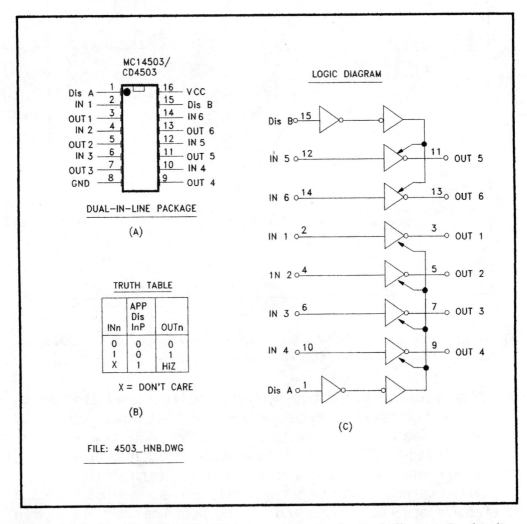

FILE: 4503_HNB.DWG

Figure 5.3.7. The CD4503/MC14503 hex non-inverting 3-state buffer. (a) Dual-in-line package. (b) The device internal logic diagram. (c) The device truth table.

Figure 5.3.7 illustrates a similar device, the MCl/CD4503. There are structural as well as logical differences between these two buffers. Note the use of the CD4503 with the four-function counter/timer. This application shows that 3-state device usage is advantageous in many situations, not simply limited to bus applications.

5.3.4. Sequential Logic — The Flip-Flop

Sequential Logic

Sequential logic differs from combinational in its ability to retain its existing status while events take place around it. An example often used is the common light switch. When we turn the switch on it remains on until turned off. We can say the switch, in a sense, remembers its latest transition.

Back in the early days of solid-state device development flip-flops acquired code letter identifications. The current CMOS data books show only three basic flip-flop types: the R-S, the D, and the J-K.

The earliest was the basic set-reset latch, which was called the R-S flip-flop. Before launching into a discussion of present-day IC flip-flops suppose we do a quick review of how these circuits function. *Figure 5.3.8* is typical discrete version of the R-S flip-flop. This circuit assumes germanium transistors which often required low base resistors connected to a negative supply to hold the transistor in the off state.

In this circuit one of the transistors would switch on at powering up the circuit, so external controls would be required to assure the required start-up state. Let's pretend Ql is on. The emitter current lifts voltage V_E above the base voltage of Q2, V_{B2}, cutting Q2 off. The collector voltage, V_{C2}, is now close to the supply, V_{CC}. A positive trigger pulse to the base of Q2 switches it to the ON state, and the two collector voltages exchange values.

Today's integrated circuits are far more complex in their construction. The base voltage in the discrete example had to achieve a level sufficient to switch the transistor, and hold it sufficiently long to assure the transition.

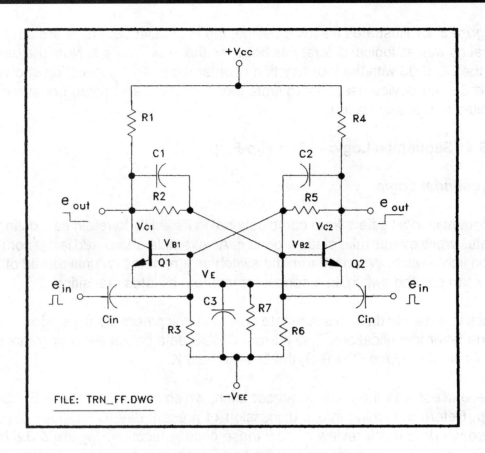

FILE: TRN_FF.DWG

Figure 5.3.8. A discrete R'S flip-flop example.

The early IC flip-flops had a similar requirement. Current flip-flops largely change state on the low-to-high transition of the input.

Figure 5.3.9 illustrates a two-input NAND gate providing the R-S latch function. The latch provides two complementary outputs. The latch state will only change under the input conditions defined by the truth table.

A nice thing about this latch is the ease with which one can be constructed from an unused pair of gates. Either NAND or NOR gates can be used, with the input logic reversed for the NOR.

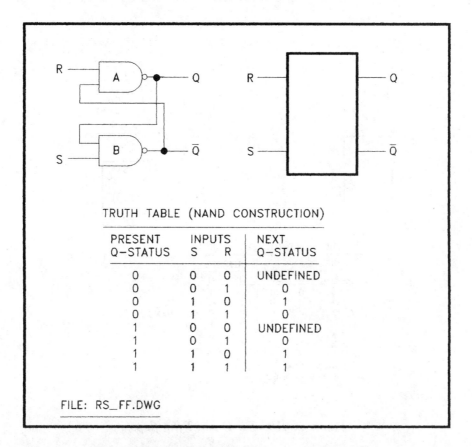

TRUTH TABLE (NAND CONSTRUCTION)

PRESENT Q-STATUS	INPUTS S	R	NEXT Q-STATUS
0	0	0	UNDEFINED
0	0	1	0
0	1	0	1
0	1	1	0
1	0	0	UNDEFINED
1	0	1	0
1	1	0	1
1	1	1	1

FILE: RS_FF.DWG

Figure 5.3.9. *The R-S flip-flop based NAND latch.*

A multitude of flip-flop types followed the introduction of the *integrated circuit*. The number has since narrowed down and two types are now able to meet most requirements: The D and the J-K.

Figure 5.3.10 shows the D type and its derivative, the T, flip-flop. This basic flip-flop can be used for virtually any requirement by tailoring the external control logic. A first look at the truth table shows the output responding to a low-to-high transition only at the clock input. We next observe how the Q output tracks the *data* (D) input: this input must be high to enable the Q transition to the 1 level. Note that the *set* (S) and *reset* (R) inputs override the clock regardless of the data.

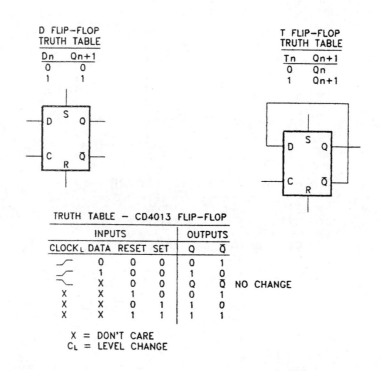

D FLIP-FLOP
TRUTH TABLE

Dn	Qn+1
0	0
1	1

T FLIP-FLOP
TRUTH TABLE

Tn	Qn+1
0	Qn
1	Qn+1

TRUTH TABLE — CD4013 FLIP-FLOP

INPUTS				OUTPUTS		
CLOCK$_L$	DATA	RESET	SET	Q	Q̄	
⌐‾	0	0	0	0	1	
⌐‾	1	0	0	1	0	
‾⌐	X	0	0	Q	Q̄	NO CHANGE
X	X	1	0	0	1	
X	X	0	1	1	0	
X	X	1	1	1	1	

X = DON'T CARE
C$_L$ = LEVEL CHANGE

Figure 5.3.10. The D and T flip-flops.

Connecting the not-Q output to the *data* (D) input results in the *toggled* (T) flip-flop. That is, the Q output will invert with each transition at the clock input. The A and B sides of the CD4013 can perform a divide-by-four function, as seen in **Figure 3.5.2.** Also note that we can tie the clock and data inputs to ground or the supply and use the set and reset as latch inputs.

There are situations in which it is advantageous for internal control of the flip-flop output. The J-K flip-flop, shown in **Figure 5.3.11**, responds only when the J and K inputs conform to the truth table requirements.

5.3.5. The IC Monostable and Timer

As with the flip-flop the IC monostable was preceded by discrete *one-shot* transistor versions. **Figure 5.3.12** illustrates a typical circuit. In this circuit

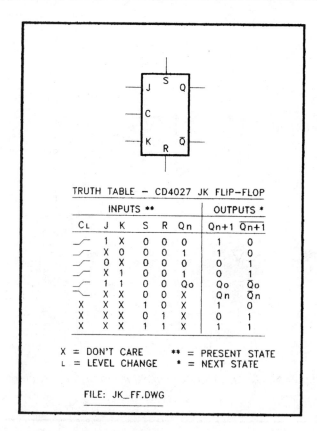

Figure 5.3.11. The J-K flip-flop.

transistor QI is held in the ON state by the bias current through resistor R3. Q2 is held OFF. Capacitor C2 is charged to very nearly the supply voltage, VCC. A negative trigger pulse to the base of QI switches this transistor OFF, diverting the QI collector current to the base of Q2, thereby ON. The reversed potential on C2 holds QI off until the time constant R3C2 washes out, allowing a return to the initial conditions.

Now, aren't we happy for the so versatile, ever so easy to use IC!

Two popular IC monostables are the 74C221 and the MCI/CD4528. *Figure 5.3.13* provides the terminal connections and operating features of the 74C221. The 74C221 is comparable to the 74HC221. These are non-

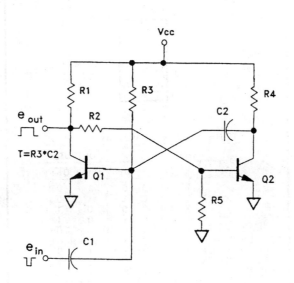

Figure 5.3.12. A discrete transistor monostable example.

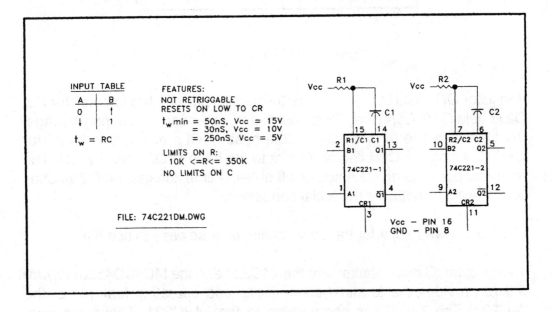

Figure 5.3.13. The 74C221 dual IC non-retriggerable monostable.

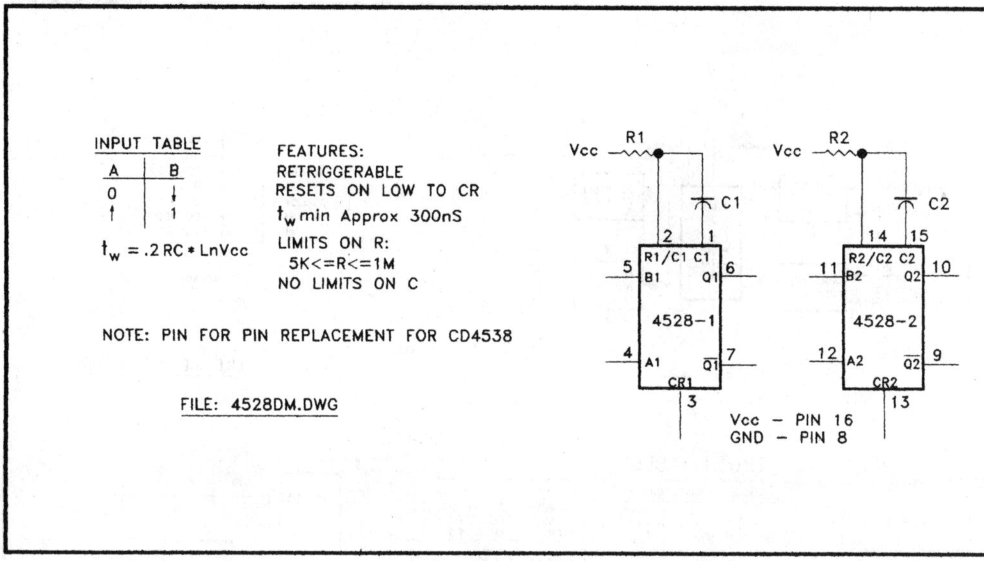
INPUT TABLE
A B
0 ↓
↑ 1
tw = .2 RC * LnVcc
FEATURES:
RETRIGGERABLE
RESETS ON LOW TO CR
tw min Approx 300nS
LIMITS ON R:
5K<=R<=1M
NO LIMITS ON C
NOTE: PIN FOR PIN REPLACEMENT FOR CD4538
FILE: 4528DM.DWG
Vcc
R1
C1
R1/C1 C1
B1 Q1
4528-1
A1 Q1
CR1
Vcc
R2
C2
R2/C2 C2
B2 Q2
4528-2
A2 Q2
CR2
Vcc - PIN 16
GND - PIN 8

Figure 5.3.14. The CD4528 dual IC retriggerable monostable.

retriggerable monostables. By this is meant that once the mono has been triggered it will not respond to further triggering until the delay from the initial trigger has timed out. By contrast, a retriggerable mono extends the delay continuously for as long as triggers are repeated. The 74HC221 provides Schmitt inputs, enabling triggers with slow rise and/or fall times.

The 74C221 is pin-for-pin compatible with the 74HC123, a dual retriggerable monostable. This monostable also features Schmitt trigger inputs.

Figure 5.3.14 describes the retriggerable CD4528 monostable. The CD4538 is a pin-for-pin compatible with tighter timing controls. The timing formula for the CD4538 is simply the product RC where R is in ohms; C in farads.

By contrast to these families, the retriggerable CD4047 is a very unique device. It can operate in either the monostable or astable, that is, free running, mode. ***Figure 5.3.15*** provides a description of the device and its operating features. The ICM7555 was introduced some years back by Intersil, now Harris Semiconductor, as a pin-for-pin CMOS enhancement of the TTL 555 timer. In addition to its low power requirements and extended voltage

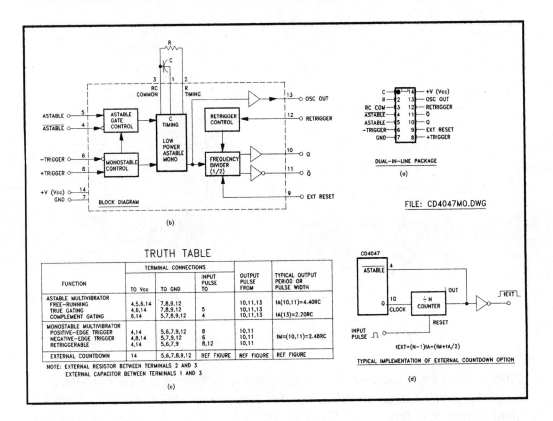

Figure 5.3.15. *The CD4047 low power monostable/astable multivibrator. (a) The DIP package. (b) Internal function diagram. (c) The truth table. (d) External countdown option diagram.*

range the ICM7555 features two astable operating modes: conformance to the equations for the 555, and a true square wave mode not available with the 555. The maximum duty cycle, D, of the 555 is 0.3. ***Figure 5.3.16*** illustrates the functions and usage of the ICM7555.

5.3.6. Latches and Decoders

We learned about the basic R-S latch earlier in section 5.3.4. The multiple latch IC enables a variety of applications, notably in the temporary storage of data for display or counting. Frequently the device includes additional logic for decoding and output driving.

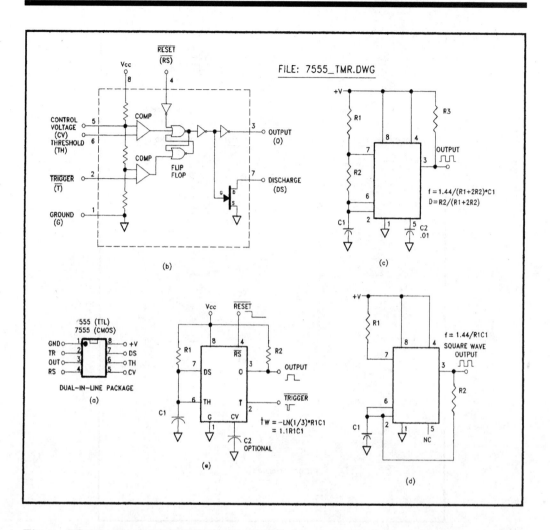

Figure 5.3.16. *The TTL555/ICM7555 CMOS timer. (a) The DIP package. (b) Internal function diagram. (c) The 7555 operating in the 555 astable mode. (d) The 7555 in the square wave astable mode. (e) The 7555 in the triggered monostable mode.*

Examples of a relatively simple latch IC are the QUAD NOR and NAND latches, CD4043/MC14043 and CD4044/MC14044, respectively. These are shown in *Figure 5.3.17*. Note the 3-state outputs. This feature facilitates sequencing a quantity of latched data onto a microprocessor bus or simply as input to a display circuit. Typical data sources requiring latching are counters, shift registers, and comparators.

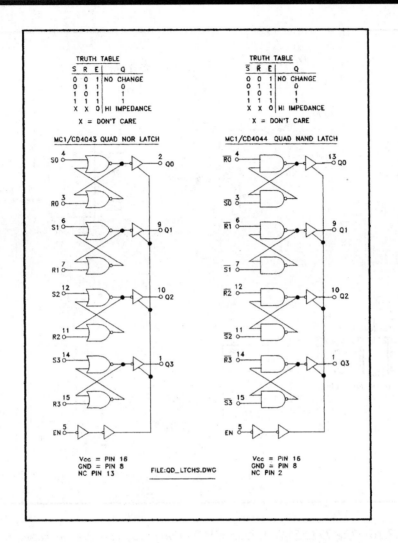

Figure 5.3.17. The MC1/CD4043 QUAD NOR, and MCI/CD4044 QUAD NAND RS latches.

The CD4511/MC14511 7-segment latch/decoder/driver is a more sophisticated device. It is often found between the output of a BCD counter and a 7-segment LED readout. Its features include sourcing up to 25 mA of segment drive, latch storage of data code, leading zero blanking, lamp test, and readout blanking on illegal input codes. It is easily implemented in a multiplexing scheme. It is very convenient to have all these functions in a single package. ***Figure 5.3.18*** shows the functioning of this IC.

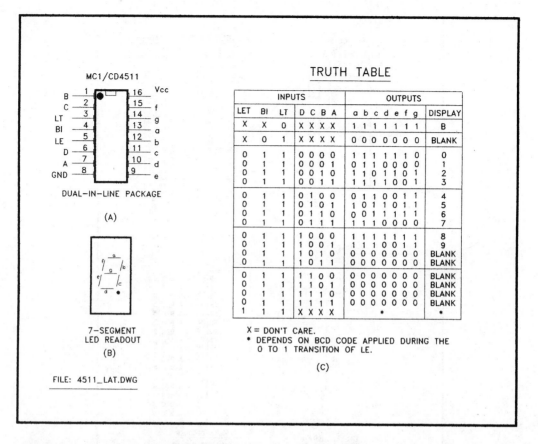

TRUTH TABLE

INPUTS							OUTPUTS							
LET	BI	LT	D	C	B	A	a	b	c	d	e	f	g	DISPLAY
X	X	0	X	X	X	X	1	1	1	1	1	1	1	B
X	0	1	X	X	X	X	0	0	0	0	0	0	0	BLANK
0	1	1	0	0	0	0	1	1	1	1	1	1	0	0
0	1	1	0	0	0	1	0	1	1	0	0	0	0	1
0	1	1	0	0	1	0	1	1	0	1	1	0	1	2
0	1	1	0	0	1	1	1	1	1	1	0	0	1	3
0	1	1	0	1	0	0	0	1	1	0	0	1	1	4
0	1	1	0	1	0	1	1	0	1	1	0	1	1	5
0	1	1	0	1	1	0	0	0	1	1	1	1	1	6
0	1	1	0	1	1	1	1	1	1	0	0	0	0	7
0	1	1	1	0	0	0	1	1	1	1	1	1	1	8
0	1	1	1	0	0	1	1	1	1	0	0	1	1	9
0	1	1	1	0	1	0	0	0	0	0	0	0	0	BLANK
0	1	1	1	0	1	1	0	0	0	0	0	0	0	BLANK
0	1	1	1	1	0	0	0	0	0	0	0	0	0	BLANK
0	1	1	1	1	0	1	0	0	0	0	0	0	0	BLANK
0	1	1	1	1	1	0	0	0	0	0	0	0	0	BLANK
0	1	1	1	1	1	1	0	0	0	0	0	0	0	BLANK
1	1	1	X	X	X	X				*				*

X = DON'T CARE.
* DEPENDS ON BCD CODE APPLIED DURING THE 0 TO 1 TRANSITION OF LE.

(c)

MC1/CD4511

DUAL-IN-LINE PACKAGE

(A)

7-SEGMENT LED READOUT

(B)

FILE: 4511_LAT.DWG

Figure 5.3.18. The MC1/CD4511 BCD-to-seven segment latch decoder/driver. (a) The dual-in-line package. (b) Standard segment designations for the display. (c) The device truth table.

The CD4514/MC14514 and CD4515/MC14515 are complementary 4-bit transparent latch/4-to-16 line decoders. The term *transparent* might be confusing. The vacuum tube counters of the forties and fifties had a counting sequence that repeated each second, with all the digits displayed while rippling through the count accumulation. That doesn't happen with these guys; the operation takes place on its own internally and we just see the end results.

Figure 5.3.19 includes a block diagram and truth table describing their operation. Note that the two functions of latching and decoding are shown separately. In use, data may be continuously input to the latch without be-

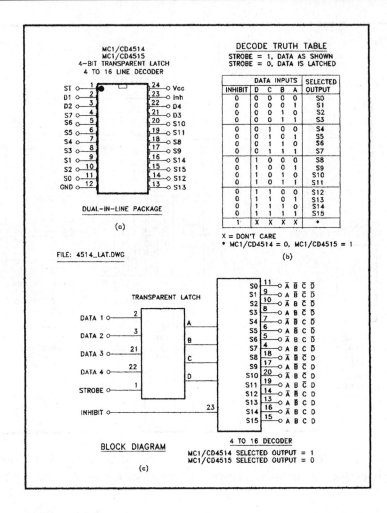

DECODE TRUTH TABLE
STROBE = 1, DATA AS SHOWN
STROBE = 0, DATA IS LATCHED

	DATA INPUTS				SELECTED
INHIBIT	D	C	B	A	OUTPUT
0	0	0	0	0	S0
0	0	0	0	1	S1
0	0	0	1	0	S2
0	0	0	1	1	S3
0	0	1	0	0	S4
0	0	1	0	1	S5
0	0	1	1	0	S6
0	0	1	1	1	S7
0	1	0	0	0	S8
0	1	0	0	1	S9
0	1	0	1	0	S10
0	1	0	1	1	S11
0	1	1	0	0	S12
0	1	1	0	1	S13
0	1	1	1	0	S14
0	1	1	1	1	S15
1	X	X	X	X	*

X = DON'T CARE
* MC1/CD4514 = 0, MC1/CD4515 = 1

(b)

FILE: 4514_LAT.DWG

Figure 5.3.19. *The MC1/CD4514 negative and MC1/CD4515 positive output 4-bit transparent latch/4-to-16 line decoders. (a) The dual-in-line package. (b) Decode truth table. (c) The device block diagram.*

ing transferred on to the decoding section. The strobe pulse sets the latches to the value the four data bits represent at that precise moment. The latch output is decoded for display as a 1 (CD4514) or a 0 (CD4515) at one of the possible 16 outputs.

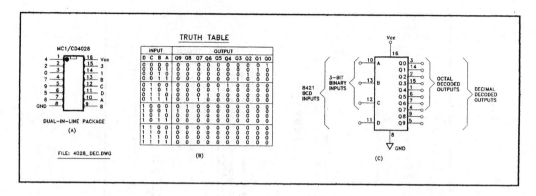

Figure 5.3.20. *The MC1/CD4028 BCD-to-decimal/binary-to-octal decoder. (a) The dual-in-line package. (b) Decide truth table. (c) The device input-output diagram.*

Data Decoders

The object of a logic decoder is to respond to data present at its input terminals with an identifying marker. This marker, which may be a high or low state on an output pin, (a logical zero or one), is available to initiate a course of action. In the absence of a latch the output will follow changes in the data.

Decoding is defined as line-to-line, such as BCD-to-decimal (ten-line) or 4-to-16 line. The CD4028/MC14028 and 54C/74C42 are 4-to-10 line decoders, capable of translating BCD input to a logical one or zero, respectively.

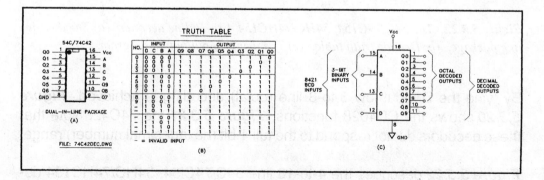

Figure 5.3.21. *The 54C/74C42, 54HC/74HC42 BCD-to-decimal/binary-to-octal decoder. (a) The dual-in-line package. (b) Decode truth table. (c) The device input/output diagram.*

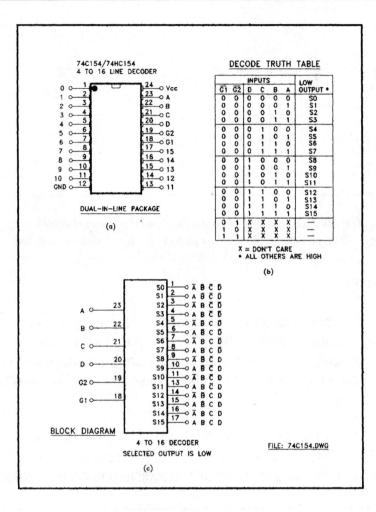

Figure 5.3.22. *The 54C/74C154, 54HC/74HC154 4-to-16 line decoder. (a) The dual-in-line package. (b) Decode truth table. (c) The device input/output diagram.*

By tying the D input low 3-to-8 line (octal) decoding is achieved. **Figure 5.3.20** shows the CD4028 functions; **Figure 5.3.21** the 74C42. Note that these decoders do not respond to the full 4-bit (hexadecimal number) range.

Figure 5.3.22 describes the 4-to-16 line 54C/74Cl54, 54HC/74HC154 decoder. These do respond to the full 4-bit range. The two enable terminals, Gl and G2, provide for a greater range of logic flexibility in their use.

Note that these ICs do not provide 3-state outputs.

5.3.7. Counters and Shift Registers

I was really happy when the first IC counters showed up; Texas Instruments TTL 7490, 7492, and 7493 led the pack if I remember correctly. It sure beat making your own from a handful of transistors, resistors and capacitors — not to mention how much more reliable in performance and the lengthened mean time between failure!

The 74C90 Ripple Counter Series

In the 1970s, National introduced their 74C00 series of CMOS ICs that were pin-for-pin replacements for the 7400/74LOO series of TTL. The 74C90

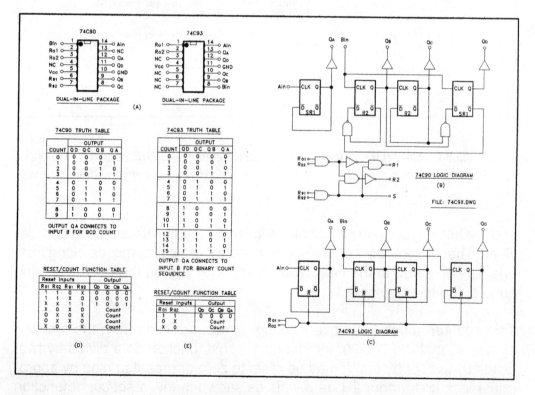

Figure 5.3.23. *The 74C90 BCD and 74C93 binary ripple counters. (a) DIP packages. (b) The 74C90 logic diagram. (c) The 74C93 logic diagram. (d) The 74C90 truth table. (e) The 74C93 truth table.*

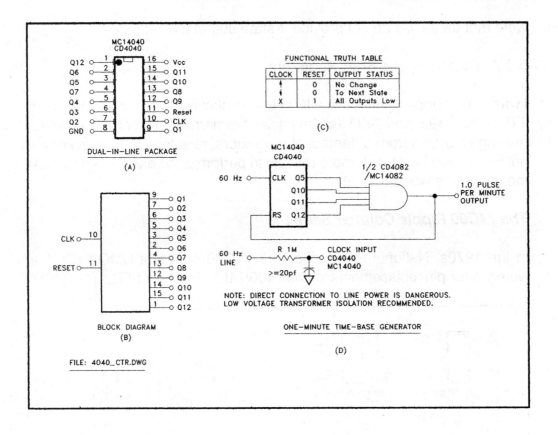

Figure 5.3.24. *The MC1/CD4040 12-bit binary counter. (a) DIP package. (b) Block diagram. (c) Functional truth table. (d) Application circuit of one-minute time-base generator.*

and 74C93 are currently available: *Figure 5.3.23* illustrates these two devices. These are *ripple counters*, which means the count moves along from one flip-flop stage to the next. The count advances on the negative-going edge of the input pulse.

This counter series is unique in bringing out terminals from the first two stages so that the 74C90, for instance, could function as a divide by two, divide by five, or by ten. It may be reset to zero or preset to nine by appropriate logic to the four 'IR" terminals as shown in the *reset/count* function table.

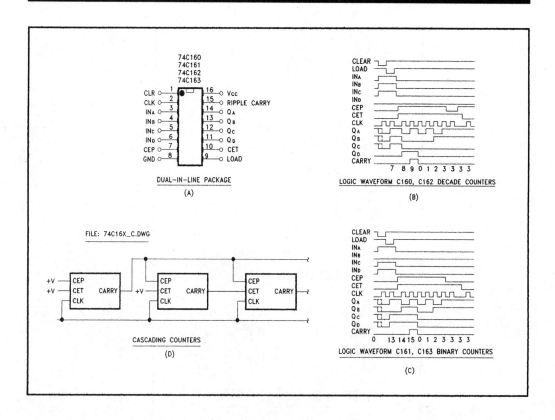

Figure 5.3.25. *The 74C160 counter series. (a) DIP package. (b) Logic waveforms, 74C160, 74C162 decade counters. (c) Logic waveforms, 74C161, 74C163 binary counters.*

Similarly, the 74C93 binary counter may be reset to zero and perform a division by two, eight, or sixteen.

The MCI/CD4040 12-Bit Binary Counter

Given the large number of available binary counters, I have had to limit the choices. I picked this one because the addition of a single 4-input AND and a 60 Hz source, which may be AC, yields a one-minute time base. Features of this counter are given in ***Figure 5.3.24***. Note from the truth table the count advances on the negative transition of the input pulse.

A similar binary counter is the 14-bit MCI/CD4020. The MCI/CD4060 is a 14-binary counter with a built-in oscillator/buffer. A quartz crystal or RC

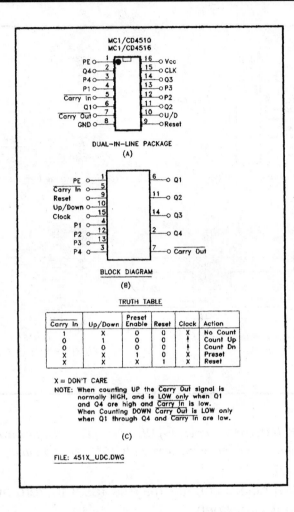

MC1/CD4510
MC1/CD4516

DUAL-IN-LINE PACKAGE
(A)

BLOCK DIAGRAM
(B)

TRUTH TABLE

Carry In	Up/Down	Preset Enable	Reset	Clock	Action
1	X	0	0	X	No Count
0	1	0	0	↑	Count Up
0	0	0	0	↑	Count Dn
X	X	1	0	X	Preset
X	X	X	1	X	Reset

X = DON'T CARE
NOTE: When counting UP the Carry Out signal is
normally HIGH, and is LOW only when Q1
and Q4 are high and Carry In is low.
When Counting DOWN Carry Out is LOW only
when Q1 through Q4 and Carry In are low.

(C)

FILE: 451X_UDC.DWG

Figure 5.3.26. The MC1/CD4510 BCD and MC1/CD4516 binary counters. (a) DIP package. (b) Block diagram. (c) Truth table.

circuit may be used. A Schmitt input allows slow rise input rise and fall pulses.

The 74Cl60 Series of BCD and Binary Counters

As with 74C90s, this series is a pin-for-pin equivalent of the TTL 74160 series. The series features are provided in *Figure 5.3.25*. Note that the 74Cl60 and 74Cl62 are BCD; the 74Cl6l and 74Cl63 are binary. The 74Cl60/ l6l feature asynchronous clear; resetting occurs whenever the reset input

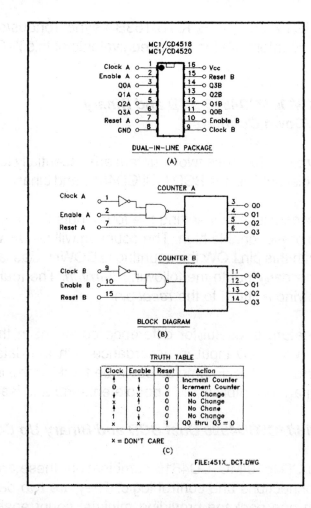

Figure 5.3.27. The MC1/CD4518 dual BCD and MC1/CD4520 dual binary counters. (a) DIP package. (b) Block diagrams. (c) Truth table.

goes LOW. The remaining two are cleared synchronously; resetting occurs on the positive clock edge following a low on the reset terminal. All four types are presentable; asynchronously or synchronously as with the reset function.

Figure 5.3.25 shows the logic waveforms for the two counter types. A diagram for cascading counters is also included. These counters include internal *carry look ahead* for faster counting and ease of cascading counters without the need for additional external logic.

The Motorola MC14160B/161B/162B/163B asynchronous/synchronous presettable 4-bit counters are functional equivalents of this 74Cl6O counter series.

The MCI/CD4510/MCI/CD4516 BCD and Binary Presettable Up/Down Counters

As seen in *Figure 5.3.26*, these two counters are essentially identical other than for the type distinctions: the BCD MCI/CD4510 and binary MCI/CD4516.

The preset is accomplished by setting the four inputs to the desired value and bringing the preset enable high. The counting will be UP with the U/D, pin 10, HIGH. With this pin LOW the counting is DOWN. Cascading is done by connecting the *carry out* to the following *carry in*. The four outputs are set LOW by applying a HIGH to the *reset* pin.

The Up/Down feature is useful for difference counting. In this, selection logic operates on the U/D input in accordance with the data sources to maintain the current status. Examples could be tracking the inputs to and the outputs from an assembly line, or people entering and leaving a room.

The MCI/CD4518/MCI/CD4520 Dual BCD and Binary Up Counters

As with the MCI/CD4510/MCI/CD4516 combination, these are identical in their terminal connections and control logic. They are two basic independent counters in one package providing minimal count enable, counting and reset functions. These are shown in *Figure 5.3.27*. Each counter is identical, independent, internally synchronous D-type flip-flop stages. The *clock* and *enable* inputs may be interchanged to clocking on either edge of the input pulse. As shown in the figure clocking occurs on the positive edge. Counters are cleared with a HIGH on the *reset* line.

The MCI/CD4017 Decade Counter

This workhorse IC is not your everyday BCD counter. Internally it consists of a 5-stage *Johnson* decade counter with code decoding. The Johnson stages contribute to high-speed, spike-free outputs. The ten decoded out-

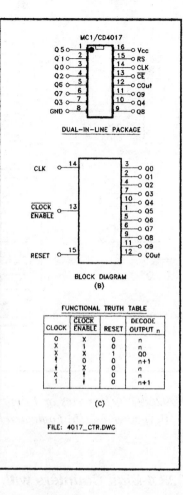

Figure 5.3.28. *The MC1/CD4017 decade counter with built-in code converter. (a) DIP package. (b) Block diagram. (c) Truth table.*

puts are in a normally LOW state, going to the HIGH level only when the appropriate input count exists. Outputs change on the positive leading edge of the clock. All outputs go LOW with a HIGH on the *reset* pin. Note that as with the MCI/CD4518/MCI/CD4520 the *clock* and *clock enable* inputs may be interchanged. A *carry out* signal enables device cascading. **Figure 5.3.28** describes packaging and truth table functions.

A useful feature I frequently employ is feeding back a selected output to the *reset* to yield a specific divide-by-N function.

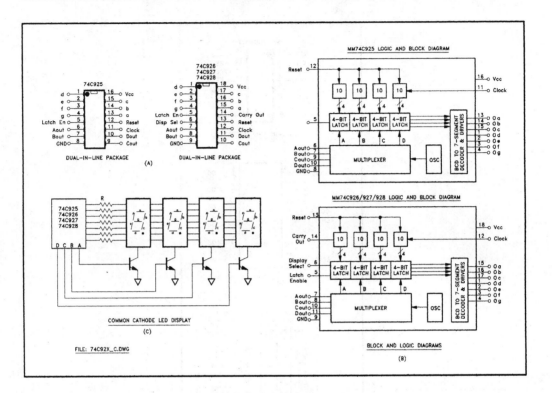

Figure 5.3.29. *The 74C925/926/927/928 series of 4-digit counters with multiplexed 7-segment output drivers. (a) DIP packages. (b) Logic and block diagrams. (c) Common cathode LED display circuit.*

The 74C925/926/927/928 4-Digit Counters with Multiplexed 7-Segment Output Drivers

This is your all-in-one counter/decoder/driver do-everything (almost) series. Be aware of one distinction from most CMOS — the power supply range is limited to 3 to 6 volts in contrast to the more usual 3 to 15 volts. An external resistor is desirable to limit internal dissipation — the drivers are capable of sourcing up to 40 mA. With all seven segments aglow, this equates to 240 milliwatts, which is significant.

Figure 5.3.29 shows the packaging, logic and block diagrams, and basic 4-LED display circuitry. Each of the four provides for clocking, data latching, and resetting. The 74C925 does not include the carry out pin required for

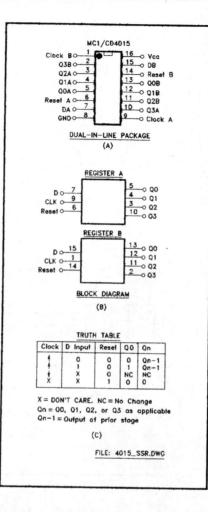

Figure 5.3.30. The MC1/CD4015 dual 4-bit static shift register. (a) DIP package. (b) Block diagram. (c) Truth table.

cascading. Note that these counters are used in the construction projects of sections 3.4 and 3.5. Consult the schematic for one of these for further insights of their usage.

The MCI/CD4015 Dual 4-Bit Static Shift Registers

As the name implies, the shift register is used for data relocation, serially or in parallel, as the device provides. A common use is the conversion of a

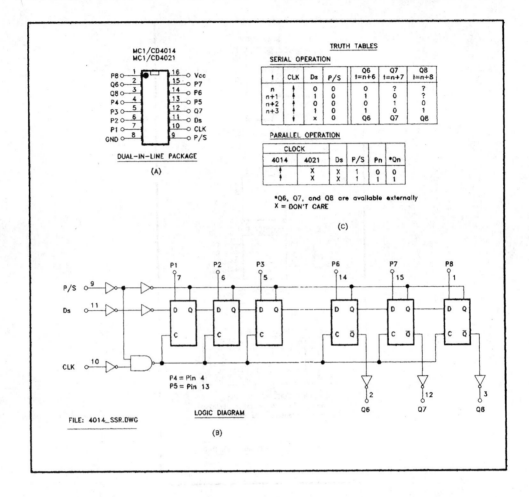

Figure 5.3.31. *The MC1/CD4014 and MC1/CD4021 8-bit static shift registers. (a) DIP package. (b) Logic diagram. (c) Truth tables.*

serial input stream to a parallel format. The converse is also possible, dependent on the device selected. This, and the MCI/CD4014/MCI/CD4021 of the following section, are included as representative of those available.

Figure 5.3.30 shows the packaging and truth table for the MCI/CD4014. The construction features D-type master-slave flip-flops. Data is shifted to the next stage on the positive going clock transition. Clearing occurs with a HIGH on the *reset* input.

The MCI/CD4014/MCI/CD4021 8-Bit Static Shift Register

These devices find application in parallel to serial data conversion with asynchronous or synchronous parallel input and serial output data queuing. Truth table functions for both serial and parallel operation are included in **Figure 5.3.31**. The figure also shows the packaging and a logic diagram. As with the MCI/CD4015 D-type flip-flops are used. Note the absence of a *reset* input.

A substantial variety of shift registers are available. I have not included them in detail as shift registers enjoy limited use in contrast to counters. The Motorola and other manufacturer data books provide details on the application of these.

5.3.8. The Digital Comparator

In section 5.2.2, we learned something of the analog comparator. So it will not surprise us to learn there exists a digital equivalent.

The first of the two types described here compare two four-bit inputs, designated A and B. The input can be binary or BCD code. Three output states are recognized: A < B, A = B, or A > B. The devices can be cascaded to

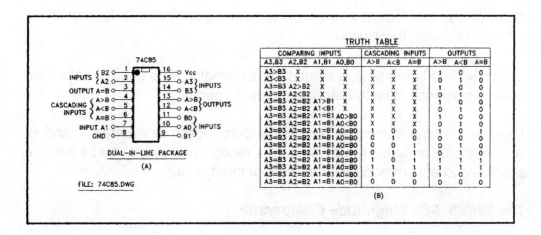

Figure 5.3.32. The 74C85 4-bit magnitude comparator. (a) DIP package. (b) Truth table.

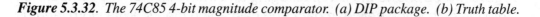

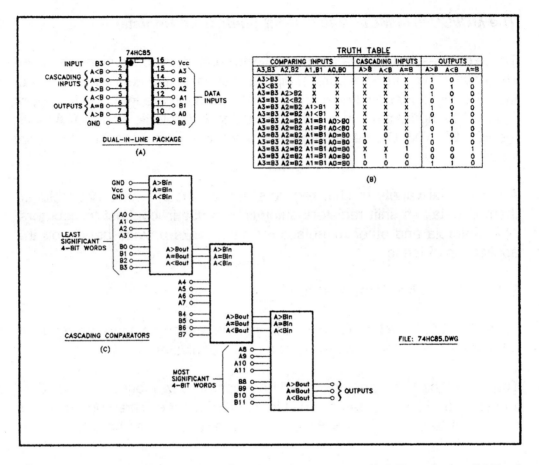

Figure 5.3.33. *The 74HC85 4-bit magnitude comparator. (a) DIP package. (b) Truth table. (c) Cascading comparators.*

accommodate larger sets of input data. External logic is used to select the desired response from the available output options.

The second type compares two 8-bit inputs, again designated A and B. With this only an exact match, A = B, is recognized. Again, these may be cascaded to accommodate larger sets of input data.

The 74C85 4-Bit Magnitude Comparator

Figure 5.3.32 shows the DIP packaging and truth table for this comparator. Note the distinction between data and cascading inputs. With word lengths

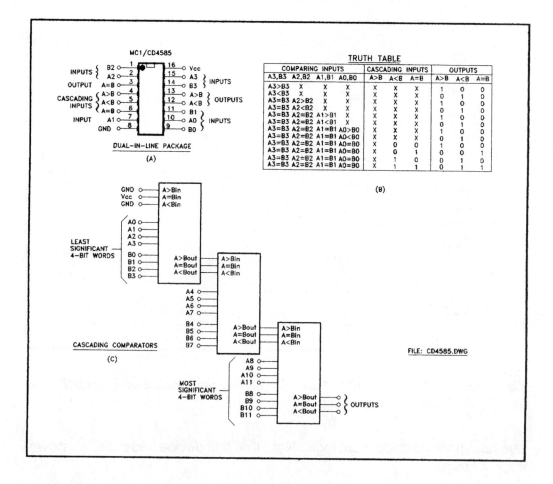

TRUTH TABLE

COMPARING INPUTS				CASCADING INPUTS			OUTPUTS		
A3,B3	A2,B2	A1,B1	A0,B0	A>B	A<B	A=B	A>B	A<B	A=B
A3>B3	X	X	X	X	X	X	1	0	0
A3<B3	X	X	X	X	X	X	0	1	0
A3=B3	A2>B2	X	X	X	X	X	1	0	0
A3=B3	A2<B2	X	X	X	X	X	0	1	0
A3=B3	A2=B2	A1>B1	X	X	X	X	1	0	0
A3=B3	A2=B2	A1<B1	X	X	X	X	0	1	0
A3=B3	A2=B2	A1=B1	A0>B0	X	X	X	1	0	0
A3=B3	A2=B2	A1=B1	A0<B0	X	X	X	0	1	0
A3=B3	A2=B2	A1=B1	A0=B0	1	0	0	1	0	0
A3=B3	A2=B2	A1=B1	A0=B0	0	0	1	0	0	1
A3=B3	A2=B2	A1=B1	A0=B0	0	1	0	0	1	0
A3=B3	A2=B2	A1=B1	A0=B0	0	1	1	0	1	1

Figure 5.3.34. *The MCI/CD4585 4-bit magnitude comparator. (a) DIP package. (b) Truth table. (c) Cascading comparators.*

exceeding four bits, cascading consists of connecting the three outputs of the least significant stage to the respective cascade inputs of the next stage. Also, the initial stage must have its three data inputs: A < 3, A = B, and A > B, connected to ground, Vcc, and ground, respectively.

The 74HC85 4-Bit Magnitude Comparator

Figure 5.3.33 describes the DIP packaging, truth table, and cascading logic for this comparator. Though similar to the 74C85 it does have differing pack-

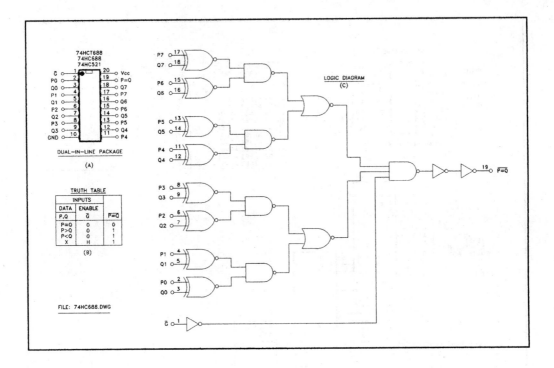

Figure 5.3.35. *The 74HC521/688/HCT688 8-bit magnitude comparator. (a) DIP package. (b) Truth table. (c) Logic diagram.*

age terminations and output logic. Note that this device has a 2 to 6V power supply range.

The MCI/CD4585 4-Bit Magnitude Comparator

While similar to the 74C85 this comparator differs somewhat in packaging terminations and cascading logic. In other respects its use duplicates the previously described devices. Details are provided in ***Figure 5.3.34***.

The 74HC521/688/HCT688 8-Bit Magnitude Comparator

These devices differ from the previously described in two ways: the ability to compare an 8-bit data input, and detection of equality only; that is, A = B. ***Figure 5.3.35*** shows the DIP packaging, truth table, and logic diagram.

The operation of the three are identical. The two '688 devices are several nS faster than the 1521; the HCT is capable of directly interfacing with TTL logic levels.

5.3.9. The A/D and D/A Converters

The Analog-To-Digital Converter (ADC)

Various techniques are in current use for conversion of an analog voltage to its digital equivalent. A direct method is the use of a reference voltage and resistance string with multiple comparators. An indirect is to employ a digital-to-analog (DAC) converter in conjunction with the A/D. A third does a conversion of the unknown voltage to an interval of time, then measures the time.

ADCs range in complexity from a basic 8-bit device to highly sophisticated units that function in a variety of data acquisition and measurement tasks. For some, an effective resolution of 18 bits is possible.

Figure 5.3.36 illustrates several possible conversion approaches. I'll touch on these briefly for an overview of their features.

Parallel (Flash) Conversion

Refer to *Figure 5.3.36* (a). In this scheme a resistor ladder generates 2n-1 voltage reference levels. Each level is connected to the non-inverting input of a two-input comparator. Note that each comparator below the peak value of the analog input is turned on. Encoding logic is required to create the output bits corresponding to the comparator states. Because the operation is conducted simultaneously on each comparator the response is very fast.

The downside is the expense, 255 comparators are required for an 8-bit converter. Their use is with digital video, radar data and similar high speed processing requirements.

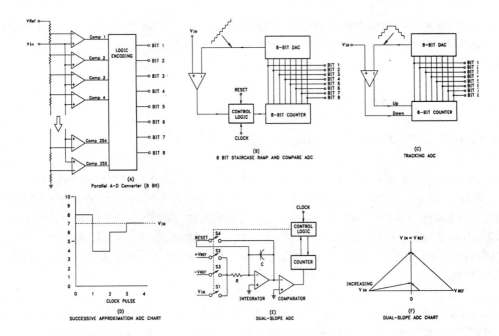

Figure 5.3.36. Some analog-to-digital conversion approaches: (a) The parallel ADC. (b) Staircase ramp and compare. (c) A tracking ADC. (d) Operation of a successive approximation ADC. (e) Dual-slope ADC logic.

Staircase Ramp and Compare Conversion

In this scheme, shown in *Figure 5.3.36* (b), the incoming analog data is compared with a staircase generated by an internal digital-to-analog converter. Each bit outputted from the ADC also increments the counter feeding the DAC.

While low in cost this method is also slow in requiring 2n-1 clock changes for a complete conversion.

The Tracking ADC

As seen in *Figure 5.3.36* (c), this scheme is similar to the staircase ramp and compare conversion. An up-down counter directly driven by the input comparator enables an immediate response to input variations. While 2n-1 clock pulses are required for the initial version subsequent conversions require but a single pulse.

The Successive Approximation Converter

Writers frequently describe this converter's operation by comparing it with a balance scale used for determining the weight of an object. In the balance scheme the sample to be weighed is placed in a sample pan suspended from the scale's balance beam. A sequence of standard weights then follows, such as beginning with a 16 gram weight, which being too heavy is replaced with an eight gram standard. This being too light, a four gram standard is added, which is close, the addition of one more gram completing the series. The operation of the scheme is shown in *Figure 5.3.36* (d).

Dual-Slope Conversion

Figure 5.3.36 (e) is a block diagram of this converter. Its operation may be a bit more difficult to follow. The diagram in the (f) portion of the figure is helpful in following the converter functioning.

Observe that there are three switches for connection to the input of an integrator, referenced to ground. A fourth switch resets the integrator. The opening and closing of the switches is performed by clocked control logic.

The initial step in the conversion cycle is the closing of switch S1, connecting the analog data to the integrator. S4 is opened and the data is integrated for n clock periods, where n is the maximum counter count. At this time the integrator output is given by $-V_{in} \bullet n \bullet T/RC$, where T is the clock period.

The signal polarity is determined by the comparator. One of the two switches, S2 or S3, is now closed to connect the reference of a polarity opposite to the input to the integrator. The effect is to "wash out" the integrated data, reducing the output to zero. This is illustrated in *Figure 5.3.36* (f). At zero, the comparator changes its output state, stopping the integration. The two count times, nT and xT, are related by $V_{in} \bullet n \bullet T/RC = V_{ref} \bullet x \bullet T/RC$, from which $x = V_{in} \bullet n/V_{ref}$. This makes x proportional to the count n and the reference voltage, both of which are constants.

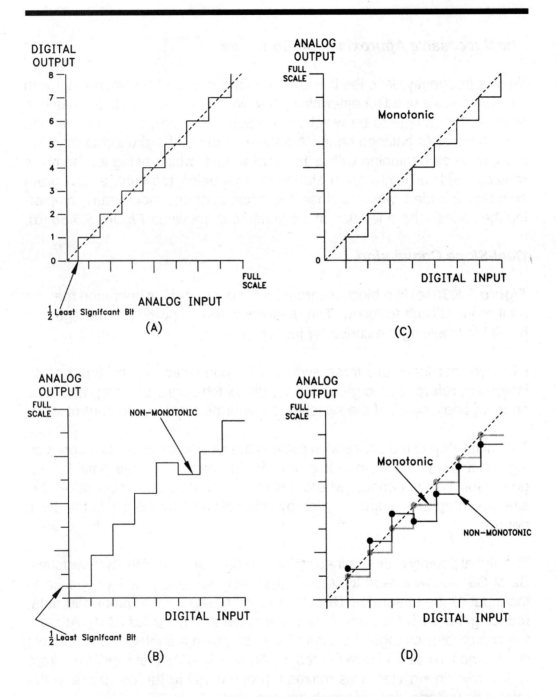

Figure 5.3.37. *ADC and DAC transfer characteristic examples: (a) Ideal 3-bit ADC. (b) Non-monotonic DAC used with an ADC. (c) Ideal 3-bit DAC. (d) Non-monotonic DAC.*

Accuracy

> *Absolute accuracy* is a measure of the total converter error, defined as the difference between the actual analog output and the expected value for a given digital input code.
>
> *Relative accuracy* is an interpretation of the converter non-linearity, defined as the difference between the actual output and the output expected based on the relative or actual full scale converter output for a given input code.
>
> Errors are expressed in LSB or percent of full-scale (fS) range.

Conversion Time

> The time for a complete measurement by the converter.

Differential Linearity (DNL)

> Adjacent digital codes should have a measured difference of one LSB (or $FS \cdot 2^{-n}$ for an n-bit converter). A deviation from the ideal difference is termed differential nonlinearity expressed in submultiples of the LSB. Errors > 1 LSB can result in non-monotonic performance in a D/A converter and missing codes in an A/D converter.

Four Quadrant

> This relates to a multiplying D/A converter. The reference voltage and the digital input may be of either polarity. The converter must observe the rules of algebraic sign multiplication.

Gain

> Converter gain is the analog scale factor relating the full scale output to its analog or reference input.

Gain Error (Full-Scale)

> *ADC:* the difference, generally expressed in terms of the LSB, between the input voltage that should produce a full-scale output code and the actual input producing that code.
>
> *DAC:* the difference between the output voltage, or current, with full-scale input code and the ideal voltage, or current, that should exist with the full scale input.

Least Significant Bit (LSB)

> The bit representing the least, or smallest, value in a binary number system. Typically this is the right-most digit. As such it represents the smallest resolvable analog change in an n-bit converter. It is equal to the FS output divided by 2^n where n = the number of ; its, that is, $LSB = FS/2^n$.

Linearity

> The deviation of the analog data from a straight line, usually specified in % or ppm of the FS range or submultiples of 1 LSB.
>
> The straight line may be either a *best straight* or *end point* construction.

Missing Codes

> A deviation exceeding 1 LSB for an incremental increase or decrease in the input voltage. With missing codes there is a numeric output value that cannot be expressed for any input.

Monotonicity

> A function having a slope whose sign does not change. As the input code of a D/A Converter (DAC) is increased in steps of 1 LSB the analog output should also increase in a stair-step fashion. When the condition is true, the converter is said to be *monotonic;* the output is a single-valued function of the input. If a decrease occurs at any input step the DAC in *non-monotonic.*

Most Significant Bit (MSB)

> The bit representing the most, or largest value in a binary number system. Typically this is the left-most digit.

Multiplying DAC

> Every DAC might be interpreted as a multiplying DAC in that the output is equal to the reference voltage times a constant determined by the input code divided by 2^n.
>
> In a two quadrant multiplying DAC the reference voltage or the input digital code can change can change the output voltage polarity. If both can change the output is of four quadrant polarity.

Offset (Zero Scale Error)

> The measured analog output when the digital input code corresponds to analog zero. May be expressed as a percentage of the FS range, or ppm, LSBs, or units of current or voltage.

Quantizing Error

> Inherent in all ADCs in that all have a limiting finite resolution. An analog input that falls between two adjacent output codes will contain an inaccuracy by up to 1/2 LSB.

Radiometric Operation

> With an absolute conversion the input voltage is compared to a fixed, stable reference. In a radiometric application the reference may have

some kind of proportional relationship with some signal sources; in which case the reference for the source should be the reference for the converter.

Resolution

The number of bits, n, that the FS range may be divided into - resolved.

Settling Time

The time required for the analog output to reach its final value following a corresponding change to the digital input code.

Switching Time (Propagation Delay)

The switching time of a D/A converter is that taken for the switching to change from the existing state to the opposite; i.e., from on to off, or vice versa. Delay and rise time, but not settling time, are included.

Temperature Coefficient (TC)

In general expressed as fractions of an LSB/°C over the rated temperature range.

Zero and Gain Adjustment

Unipolar: first adjust the DAC for output = 0 with all bits OFF. Then with all bits ON adjust the output for full scale - 1 LSB.

Bipolar: in offset binary adjust the DAC's *zero* for negative full scale with all bits OFF. Then with all bits ON adjust the output for full scale - 2 LSB.

Table 5.3.6. A glossary of ADC and DAC terminology.

Dual slope is relatively slow but offers high resolution at a low cost. A common application is in digital voltmeters.

Ideally the converter we use will output a linear code sequence of equal steps over its allowable input range. Such a transfer function is shown in *Figure 5.3.37* (a). This ideal transfer is termed *monotonic*. *Table 5.3.6* is a glossary of terms associated with A/D and D/A converters.

Let's take a moment to ponder the ADC response. We note from the plot that the analog voltage at the input varies continuously whereas the device output occurs in steps of 1 LSB. Each step is an equal fraction of the full-scale output. With the ideal converter each step is identical and there are

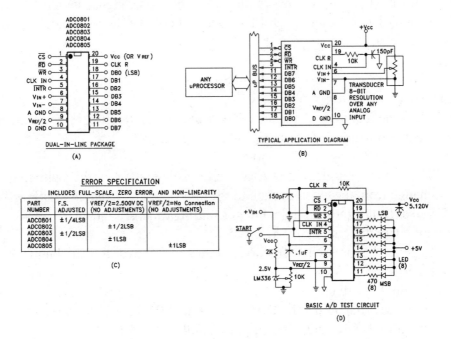

Figure 5.3.38. The ADC0801 - ADC0805 family of A/D converters. (a) D-I-L package details. (b) An application diagram. (c) Error specification table for the series. (d) A basic A-D test circuit.

no steps missing. We can draw a straight line through the plot. Note the 1/2 LSB offset at the two extremes of the plot.

Figure 5.3.37 (b) shows the non-monotonic transfer response of a DAC in use with an ADC. This in contrast to the monotonic response of *Figure 5.3.37* (c). The analog error resulting is quite apparent.

The two remaining plots of *Figure 5.3.37* compare the response of a monotonic DAC with that of a non-monotonic. I have combined the ADC and DAC plots in the single figure as the combination of ADC and DAC is fairly common. We see this in *Figure 5.3.37* where the input to a 10-bit DAC is taken from an ADC undergoing an error test.

DECODING THE DIGITAL OUTPUT LEDs

HEX	BINARY	FRACTIONAL BINARY VALUE FOR:		OUTPUT VOLTAGE CENTER VALUES V REF /2 = 2.560V DC	
		MS GROUP	LS GROUP	VMS GROUP	VLS GROUP
F	1 1 1 1	15/16	15/256	4.800	0.300
E	1 1 1 0	7/8	7/128	4.480	0.280
D	1 1 0 1	13/16	13/256	4.160	0.260
C	1 1 0 0	3/4	3/64	3.840	0.240
B	1 0 1 1	11/16	11/256	3.520	0.220
A	1 0 1 0	5/8	5/128	3.200	0.200
9	1 0 0 1	9/16	9/256	2/880	0.180
8	1 0 0 0	1/2	1/32	2/560	0.160
7	0 1 1 1	7/16	7/256	2.240	0.140
6	0 1 1 0	3/8	3/128	1.920	0.120
5	0 1 0 1	5/16	2/256	1.600	0.100
4	0 1 0 0	1/4	1/64	1/280	0.080
3	0 0 1 1	3/16	3/256	0.960	0.060
2	0 0 1 0	1/8	1/128	0.640	0.040
1	0 0 0 1	1/16	1/256	0.320	0.020
0	0 0 0 0	–	–	–	–

Table 5.3.7. Decoded values of the 8-bit ADC test LEDs.

An 8-bit ADC Example

The *National Semiconductor Corporation* offers a variety of converters of interest to experimenters. *Figure 5.3.38* illustrates the features of the ADC0801/0802/0803/0804/0805 8-bit mP compatible A/D converters. From an experimenter's point of view these are good learning devices in that they are readily available at a reasonable cost and easy to use. Although structured for microprocessor interfacing they are easily set up for independent bench top operation. An example is the test circuit of *Figure 5.3.38* (d). *Table 5.3.7* defines the decoded values of the LEDs.

Internally the family is CMOS, 8-bit converters based on successive approximation using a differential resistance arrangement. Differential analog inputs improve the common mode rejection and enable an offset analog input. The voltage reference may be adjusted to obtain the full 8-bit range from smaller analog voltage inputs.

When used with a microprocessor, they are fully compatible, appearing as memory locations or I/O ports with no interfacing required.

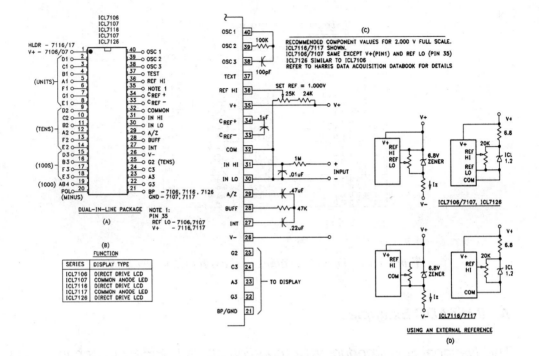

Figure 5.3.39. *The Harris ICL7106 - ICL7126 series of LCD/LED 3 1/2 digit DVM converters. (a) DIP package. (b) Function table. (c) Circuitry for 2.000 volt full scale DVM. (d) External reference schematic options.*

LCD/LED DVM Application of the Single Chip ADC

I have a Keithly 4½-digit digital voltmeter that I bought around 1979. A few months ago in one of my occasional misadventures I reversed connections to a line transformer on which I was making some measurements and wiped out the meter. After mulling over some options for a time I elected to attempt a repair on my own. In time I tracked it down to a 4½-digit ICL71C03 converter IC. These guys are packed with an amazing amount of function.

Figure 5.3.39 illustrates some features of the *Harris Semiconductor Corporation's* 3½-digit LCD/LED DVM converters. With the addition of a few resistors and capacitors a high performance digital panel can be con-

Bits	Manufacturer	Part No.	Usage
8	National	ADC0800	Radiometric, 3-state outputs
8	National	ADC0801-0805	Diff. Inputs, 3-state outputs
8	National	ADC0811	Serial I/O, 11-chan. multiplexer
8	National	ADC0816/0817	mP comp., 16-chan multiplexer
8	National	ADC0819	Serial I/O, 19-chan. multiplexer
8	National	ADC0820	mP comp., hi-spd, Trk/Hld feature
8	National	ADC0831-0834	Serial I/O w/multiplex options
8	National	ADC0833	Serial I/O w/4-chan multiplexer
8	National	ADC0852/0854	Multi. comparators w/ref. divider
10	National	ADC1001/1021	mP comp., comp w/8-bit 0801 family
10	National	ADC1005/1025	mP comp. w/2-byte interface format
12	National	ADC1205/1225	mP comp., 12-bit plus sign
12	National	ADC1210/1211	FET input, CMOS output, med. speed
-	National	ADC3511	3 1/2 digit, mP comp. BCD outputs
-	National	ADC3711	3 3/4 digit, mP comp. BCD outputs
-	National	ADD3501	3 1/2 digit DVM w/mult 7-seg output
-	National	ADD3701	3 3/4 digit DVM w/mult 7-seg output
8	Harris	ADC0802/0804	80C48, 80c50/85 bus compatible
10	Harris	CA3310/3310A	CMOS w/internal track and hold
12	Harris	HI-574A	Fast, complete w/mP interface
12	Harris	HI-674A	12mS, complete w/mP interface
12	Harris	HI-774	8mS, complete w/mP interface
4	Harris	CA3304	CMOS video speed flash conversion
6	Harris	CA3306	CMOS video speed flash conversion
8	Harris	CA3318	CMOS video speed flash conversion
-	Harris	ICL7106-7126	3 1/2 dig., LDC/LED single chip DVM
-	Harris	ICL7149	3 3/4 dig., low cost Autorange DVM
-	Harris	ICL71C03/8052	Precision 4 1/2 digit DVM
8	Analog Devices	AD570	25mS complete w/reference, clock
10	Analog Devices	AD571	25mS complete w/reference, clock
12	Analog Devices	AD368/369	Complete acq. sys., prog. gain
-	Analog Devices	AD578/579	Complete 10/12 bit, very fast
12	Analog Devices	AD678BiMOS	100nS AC/DC inp., samplg ADC

Table 5.3.8. *A listing of representative analog-to-digital converters.*

structed. (Take a look at the module described in Chapter 4, section 4.2.6.) There are cautions to be observed in this, of course. The first is with the circuit layout, in particular the analog and digital grounding. Also it is difficult to obtain high precision results with low precision, temperature sensitive, components. I have found these concerns to be particularly true with the background wiring of LCD displays.

Table 5.3.8 is a listing of selections of three manufacturer's A/D converters. I chose these three for ease of purchase by experimenters through catalog distributors. The number available from all sources far exceeds the ability to include them here.

The Digital-To-Analog Converter (DAC)

The DAC is the converse of the ADC in accepting as its input a digital value and delivering an analog output. This may be as a current or voltage, single or differential, unipolar or bipolar: the variety of available devices is just great.

As with the ADC we are concerned with linearity and the quantity defined as *monotonicity.* For convenience I included the ideal and non-ideal transfer functions for the DAC in *Figure 5.3.37*.

The DAC0800 Series

Figure 5.3.40 (b) is helpful in our understanding of how a typical DAC functions. What we see in the diagram are a comparator, transistor switches, a box into which the digital bits are input, and a simplified output derived from the switching network.

A capability of real interest is though we are given a power supply of ±15 VDC the bit inputs are TTL compatible. CMOS devices may also be used as the input source. With TTL the *threshold control*, pin 1, is grounded. With CMOS two diodes should be inserted between the pin and ground. With TTL logic the input source must be capable of current sinking but that should not present a problem with these as the maximum current is on the order of 10 µA.

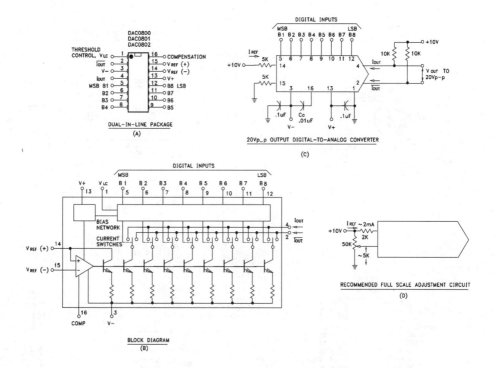

Figure 5.3.40. *The National DAC0800/01/02 digital-to-analog series. (a) DIP package. (b) Block diagram. (c) 20 Vp-p converter example. (d) Recommended full scale adjustment circuit.*

The reference voltages applied to the comparator establish the range of the output currents. When correctly adjusted a peak-to-peak range of 20 volts is obtainable.

We gain an appreciation for the high performance capability of these devices by considering some features of this series of DACs:

- 100 nS output current settling time.
- A full-scale error of ±1 LSB.
- Nonlinearity over temperature of ±0.1%.
- Full scale current drift of ±10 ppm/°C.
- High compliance of the output volts: -10 to +18.
- Complementary current outputs.
- Direct interface with TTL, CMOS, PMOS and others.

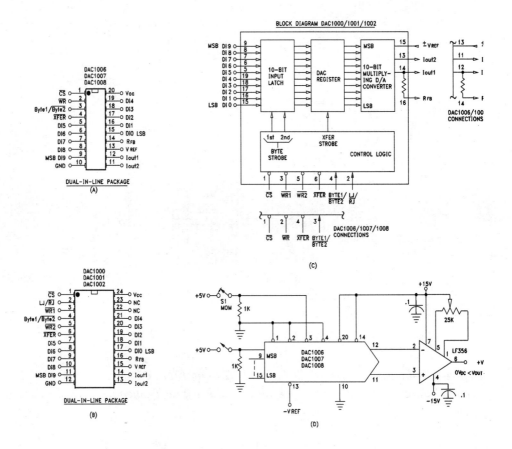

Figure 5.3.41. *The National DAC0800/01/02 digital-to-analog series. (a) DIP package,*
DAC1006/07/08. (b) DIP package, DAC1000/01/02. (c) Block diagram DAC1000/01/02.
(d) DAC1006/07/08 converter example circuit.

- Two quadrant wide range multiplying capability.
- Power range of ±4.5 to ±18 VDC.
- Power consumption 33 mW at ±5 VDC.

Res.	Manufacturer	Part No.	Usage
6	National	DAC0630/31/31A	Triple Video w/color palette
8	National	DAC0801/02/03	Hi-spd, I-output, TTL, CMOS, PMOS
8	National	DAC0806/07/08	150nS. 33mW, TTL, CMOS inputs

Res.	Manufacturer	Part No.	Usage
8	National	DAC0830/31/32	mP comp. or std alone, dbl buffered
-	National	DAC1000-1008	10,9,8-bit, mP/std alone, TTL input
10	National	DAC1020/21/22	Bin. multiplying, DTL, TTL, CMOS
12	National	DAC1220/21/22	Bin. multiplying, DTL, TTL, CMOS
12	National	DAC1208-1232	mP/std.alone
12	National	DAC1218/1219	±10v ref., multiplying 4-quadrant
8	Harris	AD7523	150nS, 8-10 bit linearity, TTL,CMOS
10	Harris	AD7520/30	500nS, 8-10 bit linearity, TTL,CMOS
12	Harris	AD7521/31	500nS, 8-10 bit linearity, TTL,CMOS
12	Harris	AD7541	+5-+15Vsup, 20mW, TTL, CMOS compat.
12	Harris	AD565/565A	±12Vsup, Hi-spd, 250nS set., 250mW,
12	Harris	HI-DAC80V	±12Vsup, Prec. Inst., CRT display
16	Harris	HI-DAC16B/16C	Hi-Fi audio, hi-res control sys.
14	Harris	ICL7134U/B	mP comp., dbl buff'd u/b-polr. inp.
16	Harris	ICL7121	mP comp., buff'd inp., bipolar
8	Analog Devices	AD557	mP comp., .2mS setl., V out
8	Analog Devices	AD7524	mP comp., .2mS setl., I out, MDAC
10	Analog Devices	AD561	.25mS setl., I out, TTL,CMOS
12	Analog Devices	AD568	.035mS setl.,10+ mA I out,TTL,CMOS
12	Analog Devices	AD668	.05mS MDAC,10+ mA I out,TTL,CMOS
12	Analog Devices	AD7548	.1mS 8-bit 4-quad MDAC, I out,TTL
14	Analog Devices	AD7534	1.5mS 8-bit 4-quad MDAC, I out,TTL
16	Analog Devices	AD569	3mS 8/16-bit 4-quad MDAC, V out,TTL

Table 5.3.9. A listing of representative digital-to-analog converters.

The DAC1000 Series

Figure 5.3.41 illustrates two subsets of the DAC1000 series: DAC1000/01/ 02 and DAC1006/07/08. Both are ten-bit devices which differ in some respects, beginning with the packaging. To show the features of both series the drawing focuses on the larger package with breakout showings of the lesser.

This is a more complex DAC than the 8-bit series. It is primarily intended for computer interfacing though that is not a strict requirement. They directly interface with most of the microprocessors in current use, appearing as a memory location or an I/O port.

One use is of multiplying applications such as digitally controlled gain blocks. (Multiplying is defined in **Table 5.3.6**.) They are also used with audio signal processing and programmable attenuators.

Double buffering is included. This appears in the figure as a latch followed by a register. The buffering can be modified to single or simple flow through.

Table 5.3.9 is a listing of selections of three manufacturers' D/A converters. As with the previous table I chose these three on the basis of ready availability. There is to vast a number available from many sources to include other than a sampling here.

A D/A Application Example

I wanted to illustrate the D/A converter while avoiding a computer interface. So I came up with the circuitry of **Figure 5.3.42**. This may look formidable, but it is pretty straightforward in most respects. The primary difficulty I experienced with it was my own ineptitude — I have a knack for leaving the power off, for instance, and wondering why the voltmeter isn't showing anything. And would you believe I connected pin four of the LM741 to ground and then spent an hour looking for the source of this awful upward offset in the output?

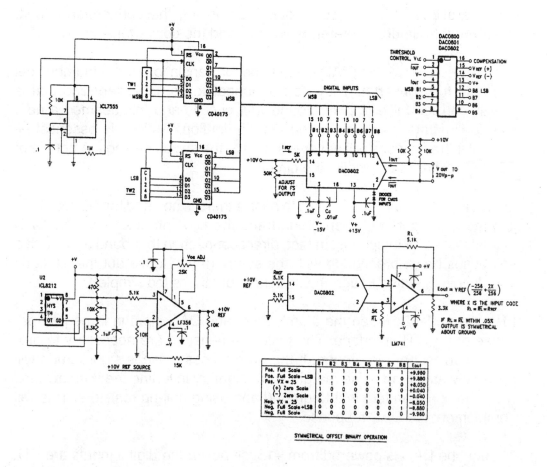

Figure 5.3.42. *A breadboard circuit using the DAC0802 converter.*

One thing I do like about the example is its use of devices from other sections of this chapter in its circuitry.

The objective of the circuit is to preset the DAC from a pair of binary coded thumbwheel switches. BNC coded switches will provide only a limited range of possibilities as the DAC ranges over the full 8-bits. Actually we only have voltages for seven bits as bit 1 (B1) is a sign bit, as seen in the table. While our attention is on the table note the two values of zero - or should we say almost zero.

There are three basic circuit blocks comprising the conversion circuit: thumbwheel-related, a reference source, and the converter.

I chose two CD40175 QUAD D flip-flops for translation of the thumbwheel status to the converter. They do not appear to be necessary and that is mostly correct. I like to play around with circuits and this seemed a neat way to illustrate operation of the D type flip-flop. A clock is essential — without it the flip-flops will never respond to the switch settings. A couple of CD4050 buffers will simplify the circuit.

In fact, if we strip the circuit down to basics there is no need for the switches, just move jumpers between ground and the five volt bus to set the DAC inputs. So we have options. In fact, direct connection to ground and the five volt bus is how I proceeded with the setup. (If you leave out the flip-flops you do this directly through the switches, but that's too simple!)

I favor a buffer between the thumbwheel and DAC input in large part because of the TTL interface. The pulldown resistors are necessary to provide a current path to ground. If the drop across these is significant it will adversely affect the noise margin. You might try inverting the thumbwheel, tying the common terminal to ground and using pull-up resistors. This will complement the DAC response.

Though the DAC is powered from ±15volt power the digital inputs are TTL compatible. The databook shows variations to accommodate interfacing several other logic families. The two diodes in series with pin 1 are for five volt CMOS sources.

There are also simpler methods for generating the ten volt reference. An advantage of the circuit shown is its stability. A ten volt zener with an op amp buffer should do it quite well. However zener diodes of this magnitude may exhibit a significant temperature drift.

Getting the DAC set up and adjusted can be frustrating. The accuracy obtainable is dependent on that of the reference voltage. The 741 op amp provides a more convenient output if we care to put the circuit to a practical use. The *National Databook* shows several alternative configurations and a number of circuit examples.

5.4. PHOTO-OPTICAL DEVICES[1]

Here we explore the behavior and some applications of a class of semiconductors that exhibit the emission of or response to infrared radiation and/or visible light. The light emitting diode (LED), introduced in the very early 1970s, became the forerunner of a wide variety of photo optical devices. Here we will gain an acquaintance with the many varieties of LEDs now available; the characteristics of phototransistors and some of the many ways in which they can be used in our projects; and features of the widely used *liquid crystal display* (LCD).

5.4.1. The Light-Emitting Diode (LED)

Background

The initial semiconductor materials were germanium and silicon. These were used for the manufacture of diodes and transistors. These materials are elements, but with the progression of the research in semiconductors attention was paid to the characteristics of materials based on combinations of elements.

I don't want to burden anyone with a dissertation on chemistry, but a bit of background can be helpful. The ninety-two naturally occurring elements are identified by a group number based on the number of valence (outer), electrons. These, which vary in number from 1 to 8, participate in molecular reactions. In general the lower the atomic number the more active the reactions. Silicon and germanium are group IV elements. The atomic number for *silicon* is 16; for *germanium* 32. Early devices were made with germanium because it was easier to process. However the properties of silicon are superior and it is the material of choice for the majority of devices at the present, except photo-optical — this because attention was directed to compounds made from group III materials, notably *gallium*, atomic number 31.

Some of these turned out to have semiconductor properties, and these are the materials from which LEDs are made. Some examples are *gallium ars-*

---

1. *Excellent references for section 5.4 are: Purdy Electronics Corporation, AND Light Emitting Diode Catalog, 1996; Purdy Electronics Corporation, AND Liquid Crystal Display Catalog, 1996; QT Optoelectronics, Optoelectronic Data Book, 1995/96. AND is a trademark of Purdy Electronics Corporation.*

Term		MKS			CGS			English		
Name	Symbol	Name	Symbol	Unit	Name	Symbol	Unit	Name	Symbol	Unit
Luminous Flux	Φ_v	Lumen	lm	lm	Lumen	lm	lm	Lumen	lm	lm
Luminous Intensity	I_v	Candela	cd	lm/sr	Candela	cd	lm/sr	Candela	cd	lm/sr
Luminance (Brightness)	L_v	Nit	nt	lm/sr-m^2	Stilb	sb	lm/sr-m^2	Candela per foot^2	cd/ft^2	lm/sr-ft^2
	L_v	Meter-Lambert (Apostilb)	mL (asb)	π – lm/ sr-m^2	Lambert	L	π – lm/ sr-cm^2	Foot-Lambert	fL	π – lm/ sr-ft^2
Illuminance (Illumination)	E_v	Lux	lx	lm/m^2	Phot	ph	lm/cm^2	Foot-Candle	fc	lm/ft^2

Radiometric System

Term		MKS		CGS	
Name	Symbol	Name	Unit	Name	Unit
Radiant Flux	Φe	Watt	W	Watt	W
Radiant Intensity	Ie	Watt per Steradian	W/sr	Watt per Steradian	W/sr
Radiance	Le	Watt per Steradian-meter^2	W/sr-m^2	Watt per Steradian-centimeter^2	W/sr-cm^2
Irradiance	Ee	Watt per meter^2	W/m^2	Watt per centimeter^2	W/cm^2
Radiant Exitence	Me	Watt per meter^2	W/m^2	Watt per centimeter^2	W/cm^2

Table 5.4.1. Photometric and radiometric terms, symbols, and units.

No. of Multiplied by Equals No. of	Foot-Lambert	Candela/m^2	Millilambert	Candela/in^2	Candela/ft^2	Stilb
Footlambert	1	0.2919	0.929	452	3.142	2,919
Candela/m^2 (nit)	3.426	1	3.183	1,550	10.76	10,000
Millilambert	1.076	0.3142	1	487	3.382	3,142
Candela/in^2	0.00221	0.000645	0.00205	1	0.00694	6.45
Candela/ft^2	0.3183	0.0929	0.2957	144	1	929
Stilb	0.00034	0.0001	0.00032	0.155	0.00108	1

Electrical	Luminosity Term	Luminosity Unit
Power (rate of energy flow)	Luminous Flux	Lumen = 1/680 watt/Luminosity Function
Power-source Output	Intensity (Power-Source)	Candela = Lumen/Steradian
Delivered Power	Luminance (Surface)	Nit = Lumen/Steradian/meter^2 = Candela/m^2
Power Transfer Efficiency	Transmittance	Transmittance Factor - 0.0 to 1.0
	Reflectance	Reflectance Factor = 0.0 to 1.0

Table 5.4.2. Conversion factors and electrical equivalents.

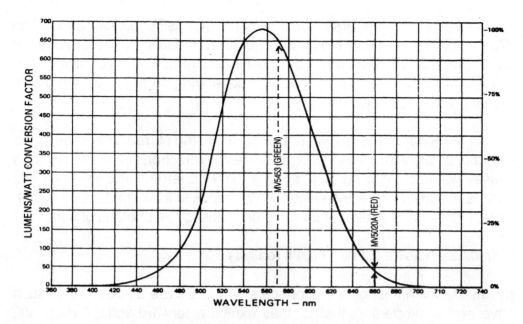

Figure 5.4.1. *The standard CIE curve used in determining the eye's efficiency at a specific wavelength and the corresponding lumens/watt at that wavelength.*

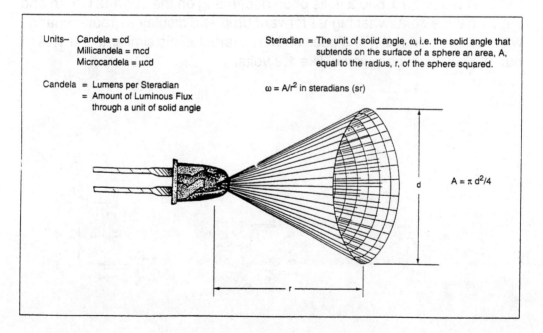

Units— Candela = cd
Millicandela = mcd
Microcandela = μcd

Candela = Lumens per Steradian
= Amount of Luminous Flux through a unit of solid angle

Steradian = The unit of solid angle, ω, i.e. the solid angle that subtends on the surface of a sphere an area, A, equal to the radius, r, of the sphere squared.

$\omega = A/r^2$ in steradians (sr)

$A = \pi d^2/4$

Figure 5.4.2. *How luminous intensity is defined as the ratio of luminous flux from a source to the solid angle subtended by the area and that source.*

enide (GaAs), *gallium arsenide phosphate* (GaAsP), *gallium phosphide* (GaP) and *indium galium aluminum phosphide* (InGaAlP). These materials have the ability to emit a narrow band of monochromatic (single wavelength) light in the visible or infrared spectrum depending on how the ingredients are selected and combined.

In their theory of operation when a forward biasing current is applied to the PN junction the electric field energy excites the electrons in the region. This causes valence electrons to jump to a higher energy state. However this state is not sustainable: the electrons drop back to their stable, lower existence, and in the process return the stored energy as photons of light.

Sounds a whole lot simpler than it really is.

Notwithstanding their optical properties, the LEDs are diodes and as such have similar properties; that is, they exhibit a forward voltage drop, VF, when conducting, and can withstand a certain level of reverse voltage, VR. Where a typical silicon diode such as the 1N914 displays a VF of 0.7 volt, a red LED will show 1.6 to 2 volts or so depending on the current. Green and yellow have a somewhat higher forward drop and display a greater change with current. Infrared (IR) diodes show a characteristic similar to the 1N914 but at a slightly higher value, like 1.2 volts.

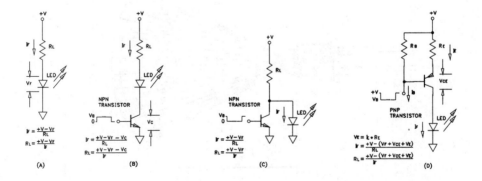

Figure 5.4.3. *LED circuit examples: (a) Continuously on. (b) Pulsed NPN transistor. (c) Inverted pulse NPN transistor. (d) Inverted pulse PNP transistor.*

Characteristic of Visible Light

Visible light is an electromagnetic radiation and as such occupies a position in the overall spectrum. Frequency is usually expressed in wavelengths in nanometers (nm) or Angstrom (Å) units rather than Hertz. Some wavelengths are: red - 660 nm; yellow - 590; green - 570; blue - 486; violet - 397. Infrared wavelengths exceed 750 nm.

The science of measuring and comparing light qualities is called *photometry*. This science has a vocabulary of its own. ***Tables 5.4.1*** and ***5.4.2*** provide detailed definitions of photometric and radiometric terminology and units. Infrared emitting sources are characterized in terms of radiometric quantities. These are simpler in that no visual corrections are needed.

Figure 5.4.1 is the standard observer curve, or "standard eyeball," established by the *Commission Internationale de l'Eclair*, called the CIE curve. In energy terms one watt of radiated energy at a given frequency is also one watt at any other. Photometric measurements require compensation for the manner in which our eyes respond to differing colors, which is to say, radiation energies. The curve represents the human eye response to a consistent level of input power.

Compare the emission energy of 7.45 lumens of the red MV5020 with the 6.49 lumens of the green MV5453. The two devices have identical input power. The red emits 180 μW of radiation at 660 nm (6600Å) whereas the green emits but 10 μW at 570 nm. Yet note the eye's very poor response to red light in contrast to green. The eye "magnifies" the green to enhance its brightness. The curve is used to calculate the lumens per watt for the radiant energy emission of a given wavelength. If we have wondered at the increasing use of blue emergency lights versus red, herein lies the answer.

The "megaphone" construction of ***Figure 5.4.2*** illustrates the distribution of luminous intensity as a function of the solid angle subtended by the device lens. Depending on the lens the LED appears as a point with a limited viewing angle or a broadly diffused source. The LED characteristics shown in the data books includes a diagram of the radiation pattern.

Visible LED Characteristics

LEDs are readily available from many manufacturers. Originally we could specify any color desired so long as it was red. That is no longer the case — LEDs are available in red, green, yellow, amber, and blue. Dual units pairing red and yellow and red and green exist. Super-bright and ultra super-bright LEDs are also available. (Are we talking LEDs or olives?)

For most needs LED circuits are relatively simple. The primary need is determination of the device current. I have found ten to twenty milliamperes to be adequate for most indoor instrumentation needs. *Figure 5.4.3* illustrates various circuit approaches. Note the limiting resistor. Because the forward drop is fixed at a low value the resistor takes up the voltage difference. Without it the LED will go *pop!*

With a five volt source I generally use a 220 ohm resistor. Device data sheets in general define the forward voltage for a current of 20 mA.

The actual illumination source is a small point in the diode PN junction. To make it useful the diode is encapsulated in a plastic lens. Three typical shapes are shown in *Figure 5.4.4* (a). Databooks and catalog drawings

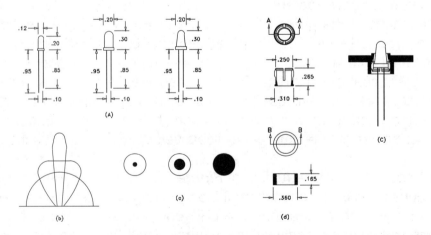

Figure 5.4.4. LED physical constructions and mounting: (a) Lens variations. (b) Varying lens effects: on-axis, broadened pattern, wide angle (flood). (c) End-on view. (d) Grommet and grooved ring mounting.

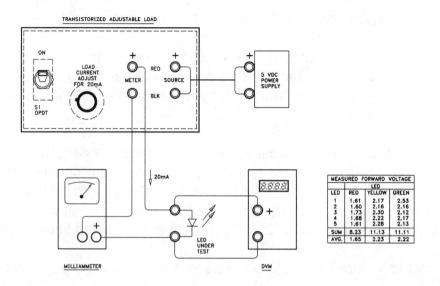

MEASURED FORWARD VOLTAGE			
	LED		
LED	RED	YELLOW	GREEN
1	1.61	2.17	2.53
2	1.60	2.16	2.16
3	1.73	2.30	2.12
4	1.68	2.22	2.17
5	1.61	2.28	2.13
SUM	8.23	11.13	11.11
AVG.	1.65	2.23	2.22

Figure 5.4.5. The measurement set up for comparing LED forward voltage drops with a 20 mA constant current.

reflect a variety of small differences — far too many to include here — that do not appear in the figure. In part these are to identify the cathode, in part for various mounting requirements.

Many LEDs have a flat on the lens side opposite the cathode lead. In practice I have found a sure way to identify this lead: I connect an ohmmeter set on the Rx10 range with the plus to the anode. On most ohmmeters the COM lead is positive. This also shows up the bad guys that are open or shorted.

Various lens shapes exist, tailored for specific viewing requirements. Three of these, illustrated in *Figure 5.4.4* (b), are for on axis, somewhat expanded off-axis, and wide angle (flood) viewing. Their end-on views appear in *Figure 5.4.4* (c)

Mounting the LED can be a challenge. The simplest method is the oldest as seen in *Figure 5.4.4* (d): a flanged, slotted grommet is inserted in the mounting hole. The LED is pushed into the grommet. A grooved locking ring is then pushed onto the low end of the grommet.

Browsing through the optoelectronics section of any catalog such as *Mouser* or *Digi-Key* will reveal a really broad selection of device packaging for almost any assembly requirement. Many are available with built-in current limiting resistors.

I made a simple test setup using the *adjustable transistor DC load* I describe in Chapter 4.2.1. This unit can be a convenient constant current source. **Figure 5.4.5** shows its use to measure the forward drop across an LED. I performed measurements on five each of red, yellow, and green LEDs to observe the drop at 20 mA. Results are tabulated in the figure.

Note that in a pinch these can be used as a simple low voltage regulator.

5.4.2. Phototransistors and Optical Coupling[1]

The Phototransistor

It turns out the PN junction is responsive to infrared and visible light. This has turned out to be a real benefit in providing electrical isolation between

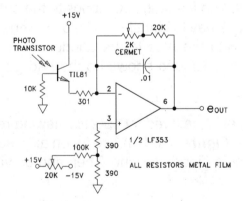

Figure 5.4.6. The author's employment of a visible light sensitive phototransistor as the input device to an analog amplifier circuit.

1. *The source for much of the content of this section is QT Optoelectronics, Optoelectronic Data Book, 1995/96.*

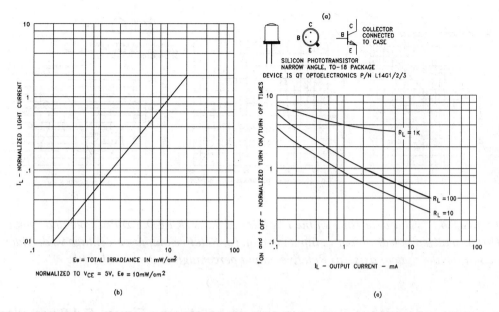

Figure 5.4.7. *Characteristics of the L14G1/2/3 silicon phototransistor. (a) Package and symbol. (b) Normalized light current vs the total irradiance. (c) Normalized turn on/turn off times vs the output current.*

circuits while maintaining communication through optical transmission. While this is largely accomplished through devices designed for optical coupling using an infrared source there is a role for the individual transistor as well. An example is an optical comparison circuit I constructed some years back.

The requirement was to compare the light transmission through two substances positioned side-by-side on a glass slide. This turned out to be a challenging design task, while exciting and fun in its doing. **Figure 5.4.6** is the circuit for one input and amplification channel. For this I constructed a light-tight structure to enclose the two phototransistors and two 6 volt lamps. Lucite rods functioned as a light pipe. The overall instrument performance successfully met a tight specification.

The TIL81 does not appear in the later Texas Instruments data book, but the L14G1/2/3 series shown in **Figure 5.4.7** appears capable of performing

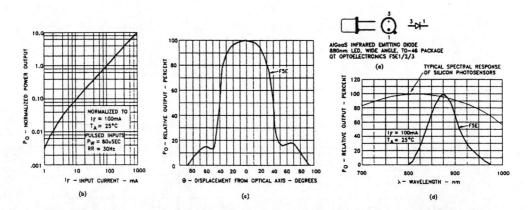

Figure 5.4.8. *Characteristics of the F5e1/2/3 AlGaAs IR LED. (a) Package and symbol. (b) Normalized power output vs input current. (c) Relative output percentage vs displacement from the optical axis. (d) Relative output percentage vs the wavelength.*

a similar function and there are sure to be others. ***Figure 5.4.8*** provides some insight into an infrared-emitting diode. The curve relating diode output to wavelength also shows a spectral response typical of silicon phototransistors.

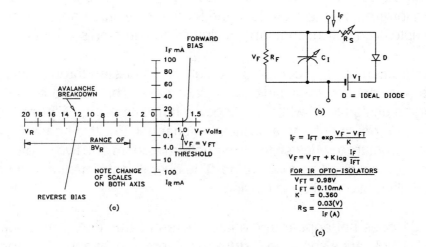

Figure 5.4.9. *(a) Typical characteristics of an IR diode emitter. (b) Equivalent circuit. (c) Circuit equations.*

Figure 5.4.10. *A selection of representative optocouplers in dual-in-line packaging. Refer to text for descriptive summaries.*

Optical Coupling[2]

Optical coupling is achieved with the combination of a light source and a photo sensitive detector. In general the light source is an infrared diode. Detectors in common use include phototransistors, Darlington pairs, SCRs, FETs, and triacs. Logic devices for both TTL and CMOS interfacing are also available.

Figure 5.4.9 illustrates the characteristics typical of an IR diode emitter. These are similar to the LED of the preceding section while differing slightly in detail. For the majority of applications the diode current is in the range of 10 to 20 mA. At higher currents attention must be given to pulse width and

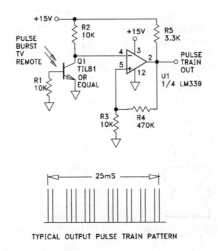

TYPICAL OUTPUT PULSE TRAIN PATTERN

Figure 5.4.11. *The author's use a phototransistor to explore the coding of his TV remote. The comparator with hysteresis provided a more visible oscilloscope trace.*

duty cycles. Of course, before making use of any device its characteristics should be reviewed.

Figure 5.4.10 is a sampling of the many optocouplers currently available. The majority require DC inputs and make use of NPN transistor or Darlington circuit outputs [*Figure 5.4.10* (a), (b), (c) and (h)] . There are exceptions of sufficient interest to devote a paragraph to their capabilities.

The MID400 [*Figure 5.4.10* (d)] is an optically isolated AC line to logic inter-face. A high gain detector circuit senses the LED current and drives the output gate to a low logic level. The device has been designed solely for use an as AC line monitor.

Some features include:

* Direct line operation from any voltage with the use of an external resistor.
* Externally adjustable time delay.
* Logic level compatibility.

Underwriters Laboratory (UL) recognition is true for the majority of couplers.

The H11F1/2/3 [*Figure 5.4.10* (e)] features a gallium-aluminum-arsenide IR LED coupled to a symmetrical bilateral silicon photodetector which performs as an electrically isolated FET. The device is designed for distortion-free control of low level AC and DC analog signals.

Some features when used as a remote variable resistor:

- $\leq$100 Ω to $\geq$ 300 MΩ.
- $\geq$ 99.9% linearity.
- $\leq$ 15 pF shunt capacity.
- $\geq$ 100 GΩ I/O isolation resistance.

Some features when used as an analog switch:

- Extremely low offset voltage.
- 60 Vp-p signal capability.
- No charge injection or latchup.
- t_{on}, $t_{off} \leq$ 15 μS.

The group represented by the 4N39, 4N40 [*Figure 5.4.10* (f)] features a gallium-arsenide IR LED optically coupled to a silicon SCR.

Some features/applications include:

- A 10A, TTL compatible, solid-state relay.
- 25W logic indicator lamp driver.
- 400V symmetrical transistor coupler.

The MOC3009/10/11 [*Figure 5.4.10* (g)] are optically isolated triac drivers. The device uses a GaAs IR LED and a light activated silicon bilateral switch which functions like a triac.

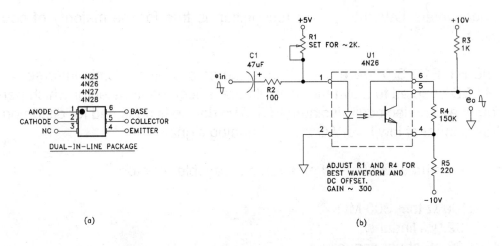

Figure 5.4.12. *The 4N25/6/7/8 phototransistor optocoupler. (a) Dual-in-line package. (b) The author's sinewave amplifier circuit.*

Some features include:

- Low input current - typically 5 mA (MOC3011).
- High isolation voltage - typically 7500V.

The H11AA1/2/3/4 [**Figure 5.4.10** (i)] is a family of devices employing two GaAs emitters connected in inverse parallel driving a single bipolar transistor. The two-diode input permits bipolar input or built-in reverse voltage protection.

The H11L1/2/3 [**Figure 5.4.10** (j)] is a family of microprocessor compatible GaAs Schmitt trigger optocouplers. The hysteresis incorporated in the Schmitt trigger enhances the noise immunity and pulse shaping. The open collector output adds to the device flexibility.

Some features include:

- High data rate, 1 MHz typical.
- Freedom from latchup and oscillation.
- Logic compatible output sinks up to 16 mA at 0.4V maximum.
- High common mode rejection ratio.
- Fast switching, 100 nS typical.

The HCPL-2503/4502, 6N136/135 [**Figure 5.4.10** (k)] are high speed optocouplers using a 700 nm GaAsP LED emitter and a high speed phototransistor. These devices contain an internal noise shield which also provides improved common mode rejection. The package design allows for a 480V isolation voltage.

Some features include:

- High speed - 1 MBit/sec
- 10 KV/µS CMR
- Current Transfer Ratio (CTR) guaranteed 0-70°C.

Note: this is the only device family described here having a reference to the CTR. The CTR, the ratio of the phototransistor's collector current, IC, to the input diode's forward current, IF, is a figure of merit. There is a temperature dependency; it is normalized to 100 percent at the standard temperature of 25° C.

For fun one day I played around with my TV remote to see somewhat of the pattern. The circuit of **Figure 5.4.11** is the result. Again I used the TIL81 which I have had on hand for some considerable years. The comparator with some hysteresis was helpful in providing good visibility with the scope traces. Someday when I have some idle time I will work up a latching arrangement so I can decode the entire range of selections.

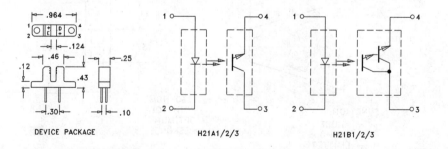

Figure 5.4.13. *The slotted optical switch. (a) Dual-in-line package. (b) H21A1/2/3 phototransistor circuit. (c) H21B1/2/3 Darlington circuit.*

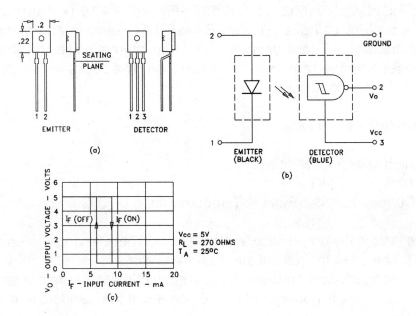

Figure 5.4.14. *The H23L1 plastic sidelooker pair. (a) Package details. (b) Electrical circuit. (c) Hysteresis characteristic.*

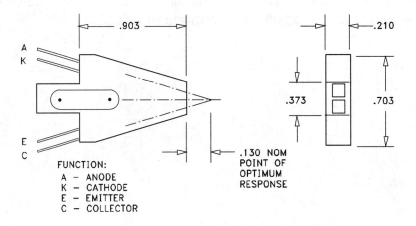

Figure 5.4.15. *The QRB1113/4 reflective object sensor package details.*

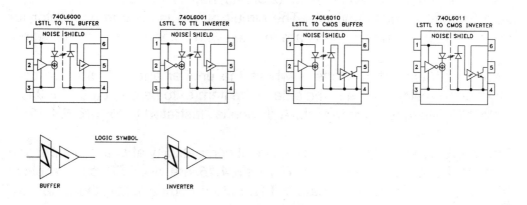

Figure 5.4.16. *The 74OL60XX high speed logic-to-logic optocoupler circuit diagrams and logic symbols.*

Nor are we restricted to ON/OFF applications of couplers. Not with all of them anyway. We observed that the H11F1/2/3 family is analog, but what about some of the others?

An answer to that question is found in *Figure 5.4.12*. Here I took a 4N26 and did some exploring with circuitry until I found an arrangement that amplified the sine wave from my function generator quite uniformly over the frequency range from about 100 Hz to better than 40 KHz. The base-emitter feedback is important; 150K worked out best with the unit I had. This kind of circuit isolation is of particular value where significant differences of ground potentials exist, such as with long circuit runs through noisy environments.

The couplers of *Figure 5.4.10* are all dual-in-line packaged. Other configurations do exist. A popular version is the slotted switch, illustrated in *Figure 5.4.13*. This is but one of several options available. The arrangement detects the presence of objects blocking the transmission between the emitter and photodetector.

Sometimes the objects are too big to slide through the slot. In this case we can go to two individual packages, as seen in *Figure 5.4.14*. The device shown, described as *plastic sidelooker pair*, has a high speed logic detector with Schmitt trigger provision. The circuit output is an open collector for flexibility in use. The two parts are marked with color dots for identification.

Another situation that exists is where the emitter and detector are constrained to reside on the same side of the item(s) to be detected. For this a reflective package does the job. A device is illustrated in **Figure 5.4.15**.

The 74OL60xx family feature logic circuit compatibility at both the input and output. The four devices shown, **Figure 5.4.16**, are all LSTTL compatible at their inputs. Two devices feature TTL output compatibility; the other two CMOS — a rather neat logic symbol.

Overall this area of opto-photo electronics encompasses a surprisingly broad array of capabilities with device packaging well adapted to the user's requirements. The major task for us may very well be the decision on which of the many available to use.

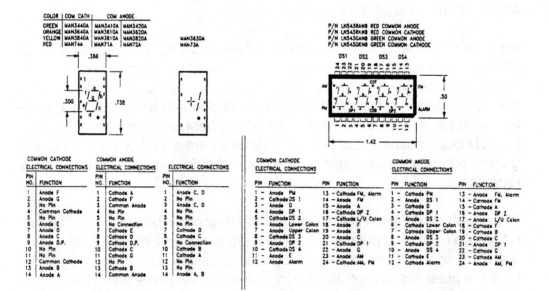

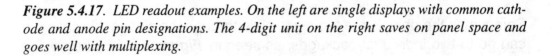

Figure 5.4.17. LED readout examples. On the left are single displays with common cathode and anode pin designations. The 4-digit unit on the right saves on panel space and goes well with multiplexing.

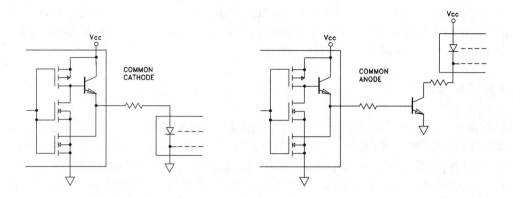

Figure 5.4.18. CMOS driver circuit examples for common cathode and common anode LED 7-segment displays.

5.4.3. LED Displays

The 7-Segment Display

Back around 1970, a small southern California company was doing quite well with a 7-segment readout using fine incandescent filaments. So far as I know those have long gone the way of the buggy whip. To paraphrase Virginia Slims, the LED display "has come a long ways, baby." As with the optocoupler there is an abundance of variety with many sizes, colors, and modular assemblies to select from.

Because of the abundance the decision on what to provide for illustration has been a challenge. For *Figure 5.4.17*, I narrowed the field to examples that appear to meet the needs of most hobbyists.

In general, the readouts come in two basic flavors: common cathode and common anode. My own preference has been common cathode, as you will notice if and when you review the count and display circuits in previous chapters. Admittedly, it is often not that more difficult to employ a common anode. A factor in favor of the common cathode is that many of the multi-function ICs tilt in that direction — the 4511/13/47 and 74C925/6/7 families, for example. These devices are similar in their output driver circuits, which feature an NPN driver transistor. We see this in *Figure 5.4.18*, affording a side-by-side comparison. Multiplexing favors the common cathode with its ground-based switch. There is merit in knowing how you plan to drive your readout before making the type decision.

The Bar Graph

Those of us who enjoy converting our LP albums to cassette tape take careful note of a flickering light array preferably maintained in the green which on particularly loud passages may display an ominous red. This practical application of the LED bar graph provides a warning whenever the recorded dB level is likely to exceed an acceptable limit.

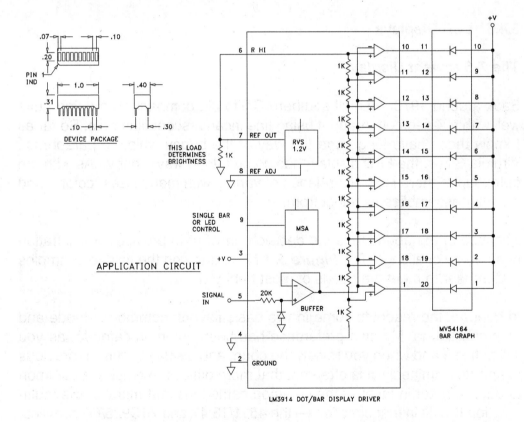

Figure 5.4.19. Bar graph package details and display driver circuit application.

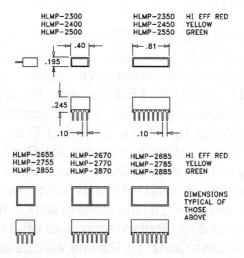

Figure 5.4.20. Some LED light bar package options. Part numbers shown are for QT Optoelectronics products.

The MV5X164[1] bar graph series is a ten-segment array of closely spaced LEDs with individual anode and cathode connections. Available are yellow and high efficiency red and green models. ***Figure 5.4.19*** illustrates the device package and its application with the LM3914 dot/bar graph display driver. Besides it use as a sound level indicator, the bar graph is a low cost, efficient way to monitor voltage levels wherever there is a need to know.

The Light Bar

This section would not be complete without reference to the several varieties of light bar currently available. The light bar is the perfect answer to our need for a highly visible yet non-intrusive indicator of something going on that we need to know about. Especially if it is going on at a distance from our point of viewing. They are available in a variety of rectangular shapes, sizes, colors, and luminous intensities. ***Figure 5.4.20*** provides a sampling of descriptions. Note that the pinouts on these are all compatible with dual-in-line, .100 inch standards.

1. *QT Optoelectronics, Optoelectronic Product Data Book, 1995/96, p 5-19.*

5.4.4. LCD Displays

I looked for a catalog source of LCD displays for a long time — one that I could actually use, that is. I did obtain some with no diagrams or information on suitable mounting hardware several years ago. This turned me off on the things and I stayed with LED displays. However, about a year ago, I made another try and discovered *Purdy Electronics Corporation*[1]. They have excellent catalogs for both LED and LCD displays along with very helpful application information. Combining the *Purdy* FE0203 3½-digit .05 inch panel display with the *Harris* ICL7106 and lots of interconnecting jumpers on a doubled-up *Super Strip* led to a working breadboard — well, mostly. The LCD is picky, and getting all the digits to work in harmony can take a bit of doing. Suppose we do a bit of exploring into what has become the mainspring of digital watches, panel DVMs and various exotic applications.

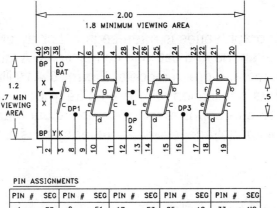

PIN ASSIGNMENTS

PIN #	SEG	PIN #	SEG	PIN #	SEG	PIN #	SEG	PIN #	SEG
1	BP	9	E1	17	E3	25	A2	33	NC
2	Y	10	D1	18	D3	26	F2	34	NC
3	K	11	C1	19	C3	27	G2	35	NC
4	NC	12	DP2	20	B3	28	L	36	NC
5	NC	13	E2	21	A3	29	B1	37	NC
6	NC	14	D2	22	F3	30	A1	38	LO
7	NC	15	C2	23	G3	31	F1	39	X
8	DP1	16	DP3	24	B2	32	G1	40	BP

FILE: LCD_DISP.DWG

Figure 5.4.21. Wiring diagram and dimensions for an AND FE LCD. Features of the device shown, the 3½ digit AND FE0203, are true for the 3 digit FE0202 as well.

1. *The symbol AND is a registered Trademark of Purdy Electronics Corporation.*

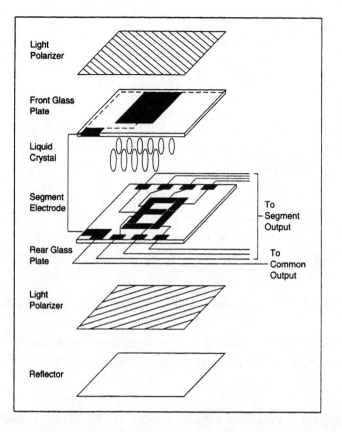

Figure 5.4.22. Construction features for the FE LCD.

Field Effect (FE) LCD Panels

Figure 5.4.21 illustrates the basic features of a connector pinned LCD. *Figure 5.4.22* shows the construction. The LCD is plugged into a socket of appropriate dimensions much as a large DIP. Unlike the DIP the construction is glass and care must be taken not to break it in the process.

There are warnings in the catalog's *Introduction, Safety and Handling Precautions* that failure to comply with certain precautions can lead to injury or even death. I am paraphrasing from the introduction on some of these.

The fluid from a damaged LCD screen is poisonous. We must not come into direct contact with it, and certainly not swallow any of it. (I have diffi-

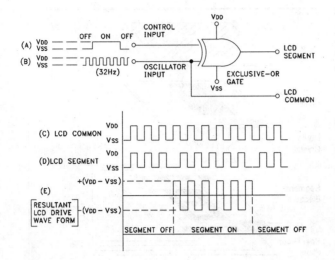

Figure 5.4.23. The input and output waveforms for an exclusive-OR drive.

culty imagining anyone doing that, but keeping these away from children makes good sense.) Should that happen, seek immediate medical assistance.

Also for certain devices (not any included in this book, however), in their CCFL and EL lines a high voltage applied to the back plane is dangerous. So if you become involved with some of these, such as graphic displays with suffixes "-EO," turn off the power before beginning work.

The polarizer and adhesive material fastening the polarizer is susceptible to damage from organic solvents and other chemicals. To clean the polarizer use a soft, lint free, cloth moistened slightly with a small amount of petroleum benzine. We should use caution when installing LCDs as the polarizer is easily scratched or damaged.

Do not apply pressure perpendicular to the plane of the glass; handle the LCD by its edges only. Also, it is recommended not to expose the device to fluorescent light or sunlight for extended periods of time.

If you feel comfortable after all that, continue on.

Integrated Circuit Drivers for FE LCDs

BCD to 4 segments	4 segments	MC1/CD4054
BCD to 7 segments	Single digit	MC1/CD4055/56, MC1/CD4543
BCD to 7 segments	Single digit	Harris/MAXIM ICM7211
BCD to 7 segments	Single digit	Siliconix DF412

Specialized ICs with LCD Driving Capabilities (AND's FE0201/03)

3-digit	LCD	A/D	Converter	Harris	CA3162
3½-digit	LCD	A/D	Converter	Harris	H17131
3½-digit	LCD	A/D	Converter	Harris/MAXIM	ICL7106
3½-digit	LCD	A/D	Converter	Harris/MAXIM	ICL7116
3½-digit	LCD	A/D	Converter	Harris/MAXIM	ICL7136
3½-digit	LCD	A/D	Converter	MAXIM	MAX139
4½-digit	LCD	A/D	Converter	Harris	ICL7129 (FE0206)
4½-digit	LCD	A/D	Converter	Harris	ICM7224 (FE0206)

Table 5.4.3. Some integrated circuits for AND FE LCDs.

We see from the construction figure, the FE LCD[2] has a front and rear glass plate. The interior sides of these are coated with a pattern of transparent and conductive material. They are mounted facing each other with a separation of a few microns. A peripheral seal of glass molding, epoxy, or the like, is applied. Both outer surfaces require light polarizing films. These may or may not be crossed, depending on the cell function. The film on the rear glass is covered with a reflective material or transreflective material. (A

APP Note #	Title
AO23	Low Cost Digital Panel Meter Designs
AO32	Understanding the Auto-Zero and Common Mode Performance of the ICL7106/7107/7109 Family
AO46	Building a Battery Operated Auto Ranging DVM with the ICL7106
AO59	Digital Panel Meter Experiments for the Hobbyist
AN9609	Overcoming Common Mode Range Issues When Using Harris Integrating Converters

Table 5.4.4. Some helpful application notes for using the ICL7106 with LCD displays.

2. *Material that follows is adapted from pages 6-2 and 6-3 of the Liquid Crystal Display Catalog, 1996.*

transreflective material reflects ambient light and transmits back light.) Images are displayed by applying a voltage between the segment and the common electrodes.

Circuit Drivers

A typical driving voltage is 5 Vrms: AND products can be driven with a range of 3 to10 Vrms. Note the "rms" implying an alternating source. The allowable frequency range is 30 to 100 Hz. A frequency below 100 Hz is recommended to conserve on power consumption.

LCDs are driven with alternating current to prevent plating of the conductive electrodes by electrolysis. Logic circuits can apply an AC symmetrical square wave. *Figure 5.4.23* is an example of such a circuit. If we recall the truth table for the exclusive-OR, we can understand the simplicity of this approach. A listing of some other ICs for use with FE LCD circuits is provided in *Table 5.4.3*.

Although I have done some "playing around" — the most truthful way of stating it — with some of the CMOS ICs, I tilt in favor of the specialized ICs, such as the ICL7106.

An area of confusion to me was the distinction between *common* and the *backplane* connections. The *Harris Corporation* offers several application notes (*Table 5.4.4*) that are helpful in providing answers to questions of this sort. The *backplane, common,* and *in lo* terminals of the ICL7106, for example, are distinct. In general, unused segments on the LCD display should be tied to the backplane pin. Connections to ground and common may present a problem. Residue on our circuit board can result in detrimental leakage paths. *Figure 5.4.24* illustrates a simple 3½-digit A/D converter circuit using the ICL7106.

We can keep in mind the DVM module of Chapter 4 waiting in the wings, as it were.

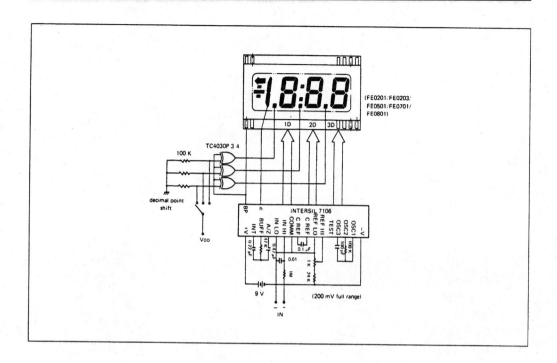

Figure 5.4.24. *An application circuit employing the ICL7106 with the AND FE series of LCDs. Note the exclusive-OR circuit for decimal point display.*

APPENDICES

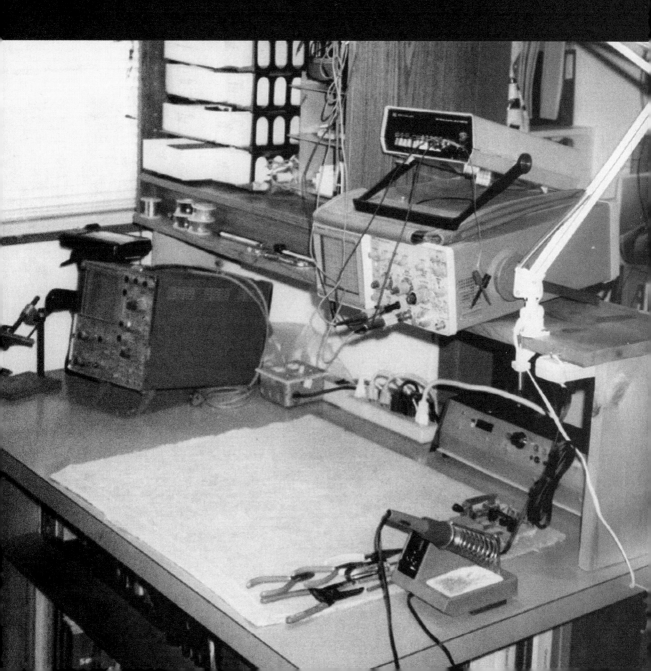

Appendix A
Catalog Mail/Telephone Order Sources

1. Altex Electronics
   11342 N IH 35
   San Antonio, TX 78233
   (800) 531-5369
   FAX: (210) 637-3264

2. Digi-Key Corporation
   701 Brooks Ave. South
   P.O. Box 677
   Thief River Falls, MN 56701-0677
   (800) 344-4539
   FAX: (218) 681-3380

3. Jameco Electronic Products
   1355 Shoreway Road
   Belmont, CA 94002-4100
   (800) 831-4242
   FAX: (800) 237-6948

4. JDR Microdevices
   1850 South 10th St.
   San Jose, CA 95112-4108
   (800) 538-5000
   FAX: (800) 237-6948

5. Mouser Electronics
   National Circulation Center
   2401 HWY 287 N.
   Mansfield, TX 76063-4827
   (800) 346-6873

6. Newark Electronics
   Chicago, IL 60640
   Information number
   (800) 463-9275

APPENDIX B
SEMICONDUCTOR MANUFACTURER LITERATURE SOURCES

Analog Devices
One Technology Way
P. O. Box 9106
Norwood MA 02062-9106
Tel (617) 329-4700
(800) 262-5643

Burr-Brown
6730 Tucson Blvd.
Tucson AZ 85706
Tel (520) 746-1111
(800) 548-6132

Crystal Semiconductor
PO Box 17847
Austin TX 78760
Tel (800) 888-5016

Dallas Semiconductor
4401 S. Beltwood Parkway
Dallas TX 75244
Tel (972) 450-0400
FAX (972) 371-4470
(800) 336-6933

Datel
11 Cabot Blvd.
Mansfield MA 02048
Tel (508) 339-3000
(800) 233-2765

Diodes Inc.
3050 E. Hillcrest Drive
Westlake Village, CA 91362
Tel (805) 446-4800

Exar Corporation
PO Box 49007
San Jose CA 95161
Tel (408) 732-7970

Hewlett-Packard Manuals
19310 Pruneridge Ave.
Cupertino CA 95014
Tel (800) 752-0900

Harris Semiconductor Literature
Dept.
PO Box 883, MS CB1-28
Melbourne FL 32902
Tel (407) 724-3739
FAX (407) 724-3937

Intel Corporation
2200 Mission College Blvd
Santa Clara CA 95052
(800) 548-4725

International Rectifier
233 Kansas St.
El Segundo CA 90245
Tel (310) 252-7106

Linear Technology
1630 McCarthy Blvd
Milpitas CA 95035
Tel (408) 432-1900

Maxim
120 San Gabriel Dr.
Sunnyvale CA 94086
Tel (800) 998-8800
FAX (408) 737-7194

Microsemi
580 Pleasant St.
Watertown MA 02172
Tel (617) 926-0404

Motorola Literature Distribution
Center
PO Box 20912
Phoenix AZ
(800) 441-2447

National Semiconductor
Dept. CRG
1111 W. Bardin Rd.
Arlington, TX 76017
Tel (800) 272-9959

NJR Corporation (JRC)
440 East Middlefield Road
Mountain View, CA 94043
Tel (415) 961-3901
FAX (415) 969-1409

Philips/Signetics
PO Box 3409
Sunnyvale CA 94088
Tel (800) 234-7381

Purdy Electronics Corporation
720 Palomar Avenue
Sunnyvale, CA 94086
Tel (408) 523-8224
FAX (408) 733-1287

QT Optoelectronics
610 North Mary Avenue
Sunnyvale, CA 94086
Tel (800) 533-6786

Samsung Semiconductor
3655 N. First St.
San Jose CA 95134
Tel (800) 446-2760

SGS-Thomson
1000 E. Bell Road
Phoenix AZ 85022
Tel (602) 485-6259

Siliconix
2201 Laurelwood Road
Santa Clara CA 95054
Tel (800) 554-5565
FAX (408) 727-5414

Teccor Electronics, Inc.
A Siebe Company
1801 Hurd Drive
Irving, Texas 75038-4385
Tel (927) 580-1515
FAX (927) 550-1309

TelCom Semiconductor
1300 Terra Bella Ave.
Mountain View CA 94043
Tel (415) 968-9252

Texas Instruments
PO Box 809066
Dallas TX 75380
Tel (800) 336-5236

Appendix C

Miscellaneous Parts, Test Equipment Sources

1. Alfa Electronics
   PO Box 8089
   Princeton, NJ 08543
   Tel (609) 897-0168
   FAX (609) 897-0206
   Orders (800) 526-2532
   High quality test equipment

2. All Electronics Corporation
   P. O. Box 567
   Van Nuys, CA 91408-0567
   Tel (818) 904-0524
   FAX (818) 781-2653
   e-mail allcorp@allcorp.com
   Misc. parts, free 64 page catalog

3. Alltronics
   2300 Zanker Road
   San Jose, CA 95131-1114
   Tel (408) 943-9773
   FAX (408) 943-9776
   Web page:
   http://www.alltronics.com
   On-line ordering:
   (408) 943-0622-14400-N-1
   24 hours
   Catalog $2.00, refundable
   first order

4. Brigar Electronics
   7-9 Alice St.
   Binghamton, NY 13904
   Tel (607) 723-3111
   FAX (607) 723-5202
   e-mail: BRIGAR2@aol.com
   Web page:
   http://members.aol.
   com/brigar2/brigar.html

Serving industries, schools, hobbyists, experimenters, consumers

5. C&H Sales
    2176 E. Colorado Blvd.
    Pasadena, CA 91107
    Tel (800) 325-9465
    FAX (818) 796-4875
    Free 128 page catalog. Power supplies, electronics, optical

6. Capital Electronics, Inc.
    303 Sherman St.
    Ackley, Iowa 50601
    Tel (515) 847-3888
    FAX (515) 847-3889
    BBS (515) 847-3890
    e-mail: sales@capital-elec.com
    Web page: http://www.
    capital-elec.com

7. Circuit Specialists, Inc.
    220 S. Country Club #2
    Mesa, AZ 85210
    Tel (602) 464-2485
    FAX (602) 464-5824
    Orders (800) 528-1417,
    Catalog request ext. 5
    Web page: www.cir.com

8. Contact East
    335 Willow St.
    N. Andover, MA 01845-5995
    Tel (508) 682-2000, free catalog number
    FAX (800) 225-5317

Orders (800) 225-5334
Web page:
www.contacteast.com
Tools, test equipment

9. Danbar Sales Company
    14455 N. 79th St., Unit C,
    Scottsdale, AZ 85260
    Tel (602) 483-6202
    FAX (602) 483-6403
    Quality reconditioned test equipment

10. DC Electronics
    P.O. Box 3203
    Scottsdale, AZ 85271-3203
    Tel (602) 945-7736
    FAX (602) 994-1707
    Orders (800) 467-7736,
    (800) 423-0070
    Miscellaneous parts

11. Debco Electronics, Inc.
    4025 Edwards Rd.
    Cincinnati, OH 45209
    Tel (800) 423-4499
    Web page: www.debco.com
    Miscellaneous parts, Electronic Experimenter's Journal

12. Gateway Electronics, Inc.
    8123 Page Blvd.
    St. Louis, MO 63130
    Tel (314) 427-6116
    9222 Chesapeake Dr.
    San Diego, CA 92123
    Tel (619) 279-6802

2525 Federal Blvd.
Denver, CO 80211
Tel (303) 458-5444
FAX (314) 427-3147
Orders (800) 669-5810
Web page:
www.gatewayelex.com
Miscellaneous parts

13. Ham Radio Outlet
(800) 559-7388
Radios, scanners,
hobbyist products

14. HSC Electronics Supply
3500 Ryder St.
Santa Clara, CA 95051
Tel (408) 732-1573
4837 Amber Ln.
Sacramento, CA 95841
Tel (916) 338-2545
6819 Redwood Dr.
Cotati, CA 94931
Tel (707) 792-2277
Orders (800) 442-5833
FAX (408) 732-6428
Web page: www.halted.com
Internet:
HSC-Info@hsc>cue.com
"...Potpourri of high tech goodies
for the techno-tinkerer"

15. Jensen Tools, Inc.
7815 S. 46th St.
Phoenix, AZ 85044-5399
Tel (602) 968-6231

FAX (602) 438-1690,
(800) 366-9662
Orders (800) 426-1194
Web page:
www.jensentools.com
Tools, test equipment

16. MEI/Micro Center
1100 Steelwood Rd.
Columbus, OH 43212
Tel (800) 634-3478
FAX (614) 486-6417
Computer related supplies,
equipment

17. MCM Electronics
650 Congress Park Dr.
Centerville, OH 45459-4072
Tel (800) 543-4330
FAX (937) 434-6959
Miscellaneous parts, equipment

18. MWK Industries
1269 W. Pomona
Corona, CA 91720
Tel (909) 278-0563
FAX (909) 278-4887
Sales/Service: (800) 356-7714
E-mail: mkenny1989@aol.com
Web site:
www.mwkindustries.com
Laser and optical related
equipment

19. Ramsey Electronics, Inc.
793 Canning Pkwy
Victor, NY 14564

Tel (716) 924-4560
Orders (800) 446-2295
Electronic hobby and amateur
radio kits

20. R&S Surplus
    1050 E. Cypress St.
    Covina, CA 91724
    Tel (818) 957-0846
    FAX (818) 967-1999
    Surplus test equipment

21. Surplus Sales of Nebraska
    1502 Jones St.
    Omaha, NE 68102
    Tel (402) 346-4750
    FAX (402) 346-2939
    Orders (800) 244-4567
    E-mail:
    grinnell@surplussales.com
    400 page catalog of parts, $5.00,
    refundable first order

22. Tucker Electronics
    1717 Reserve St.
    Garland, TX 75042
    Tel (800) 527-4642, ext. 0
    Test equipment

23. Western Test Systems
    530 Compton St., Unit C,
    Broomfield, CO 80020
    Tel (303) 438-9662
    FAX (303) 438-9685
    Orders (800) 538-1493
    Test equipment, 10 day
    inspection, 90 day warranty

INDEX